George J Manson

Smoking

A world of curious facts, queer fancies, and lively anecdotes about pipes,

tobacco and cigars

George J Manson

Smoking

A world of curious facts, queer fancies, and lively anecdotes about pipes, tobacco and cigars

ISBN/EAN: 9783337325527

Printed in Europe, USA, Canada, Australia, Japan

Cover: Foto ©berggeist007 / pixelio.de

More available books at **www.hansebooks.com**

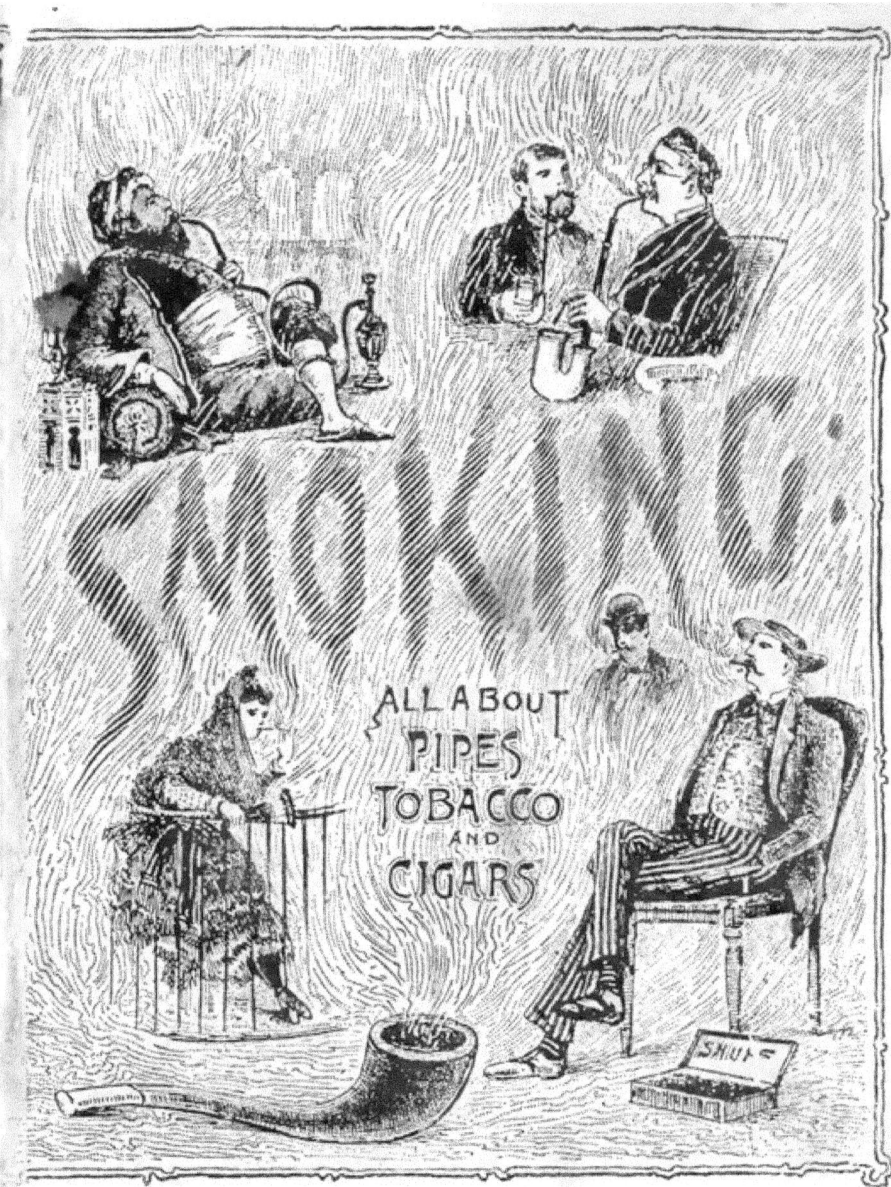

SMOKING

ALL ABOUT
PIPES
TOBACCO
AND
CIGARS

SMOKING:

A World of Curious Facts, Queer Fancies, and Lively Anecdotes about

PIPES,

TOBACCO,

AND CIGARS.

PUBLISHED BY
UNION BOOK CO.,
BROOKLYN, N. Y.

I.

What Tobacco is and where it Comes from—Its Manufacture a Government Monopoly in France—Anecdote: "Your coat-tail is on fire"—Smoking among the Chinese—The Indians Looked upon Tobacco as a Gift of the Gods—"The Calumet of Peace"—Tobacco in Europe—Its Great Value as a Medicine—Various Cures Effected by its Use—Early Efforts to Suppress the Weed—A Moral Lesson Drawn from Smoke: "Thus Think and Smoke Tobacco."

Botanically speaking there are forty varieties of the tobacco plant, growing to an altitude of from three to fifteen feet from the ground. It is cultivated in Germany, Holland, European Turkey, China, East Indies, Persia, parts of Asiatic Turkey, Cuba, Hayti, Porto Rico, Brazil, and, in our own country, in the States of Virginia, Maryland, Kentucky and Connecticut. European tobacco is not as strong as that grown in America. The tobacco grown in Germany, for instance, may be smoked continually without any bad effects; if the lover of the weed used the same amount of the American variety the effect would be very disagreeable, even dangerous.

Although the method of cultivation is the same in all countries, the differences that exist in the taste and perfume of tobacco come from the natural richness of the soil and the excellence of the temperature. The best tobacco is grown in Cuba, Mexico, Brazil, and, above all, in the United States, where the soil is fertile and the sky pure and full of sun. After Cuba, the choicest tobacco comes from Virginia, Borneo, Ceylon, and the Philippine Islands. In Asia, and principally in Persia, the cultivation is carried on extensively. As for the Turkish tobacco, it is extremely aromatic. The best brands come from Roumelia, Syria, Nomadan, Karamania, and the borders of the Persian Gulf. China furnishes a straw-yellow tobacco, which is smoked a good deal in England. Japan, Cochin China, India, and the Tonkin produce only mediocre varieties. Burmah is more favored. At Manila the cultivation is more and more important; Manila cigars are sent all over the extreme Orient. Holland has valuable tobacco lands at Java and Sumatra. The products are sold at Amsterdam, and are used throughout Europe as wrappers for costly cigars.

The United States is the most productive country in the world, and at least half of its production is exported. Mexico and

Brazil furnish very aromatic tobaccos; that of Brazil is the most combustible in the world. A great variety of species is also cultivated throughout Europe, but these are generally of very ordinary quality, and are consumed at home. England is the only country where tobacco is not grown. The German tobaccos are mostly cultivated on the borders of the Rhine, at Baden and at Mayence. They are fresh and light, but of poor flavor.

The weed was introduced into France by a Frenchman named Jean Nicot, and from him the botanical name, nicotine, is derived. The manufacture of tobacco was free in 1621, and for a long time proved a profitable business. Napoleon was attracted at a ball in the Tuileries by a lady gorgeously dressed and bedecked with many diamonds and jewels. "Who is that princess?" he inquired. When he was told that she was only a tobacco manufacturer's wife he at once resolved to take charge of this means of acquiring wealth.

In France, tobacco, being a Government monopoly, can be grown only by permission. The cultivators have the choice of selling their crops to the Government or of exporting them. No Frenchman, other than an authorized cultivator, can have tobacco leaves in his possession, and no one can keep a stock of manufactured tobacco other than that supplied by the Government, and this stock cannot exceed twenty pounds. Tobacco is now cultivated in twenty-two departments—the Nord, Pas-de-Calais, Ille-et-Vilaine, Gironde, Dordogne, Corrèz, Lot et-Garonne, Lot, Landes, Hautes-Pyrénées, Vaucluse, Bouches-du-Rhône, Var, Alpes-Maritimes, Isère, Savoie, Haute-Savoie, Puy-de-Dôme, Haute-Saône, Vosges, Meuse, Meurthe-et-Moselle. The tobaccos grown in the Nord and the Pas-de-Calais are the most charged with nicotine. Those of the Lot and the Lot-et-Garonne are the best. There exist in France nineteen tobacco manufactories, of which two are at Paris. The ordinary caporal, or, as it is officially called, *scaferlati* tobacco, is sold at $1.25 a pound, and the superior *scaferlati* at $1.60 a pound. This tobacco is put up in small packages of different colored paper. The monopoly yields the Government nearly $50,000,000 annually.

The crop alluded to above represents the home production, but the government imports a great deal of tobacco in leaf and manufactures it in France.

All the tobacco stores in France belong to the State. There are over 40,000 of them. The State does not sell tobacco at retail except in three stores ; the others are let to widows of officers, government officials, and sometimes to the widows of Senators, Deputies, and Prefects. They take the place of pensions. If the Government grants a pension to the wife of some man who has died in the service of his country, that generally means that she gets a tobacco store, or bureau, as it is called. As the social position of the pensioners will not allow them to run the bureaus directly, they let them. The dealer is allowed ten per cent. profit by the Government, and is prohibited from selling any tobacco except that supplied and priced by the State. Neither must they make cigarettes out of the Government tobacco. Every cigarette must bear the official stamp.

There is a good story of an Englishman and a Frenchman who were travelling together in a diligence, both smoking. Monsieur did all in his power to draw his phlegmatic fellow-passenger into conversation, but to no purpose. At last, with a superabundance of politeness he apologized for drawing his attention to the fact that the ash of his cigar had fallen on his waistcoat and that a spark was endangering his neckerchief. The Englishman, now thoroughly aroused, exclaimed, "Why the devil can't you let me alone? Your coat-tail has been on fire for the last ten minutes, but I didn't bother you about it!"

The smoking of tobacco is of great antiquity among the Chinese because on their very old ornaments pictures can be seen of the same tobacco pipes now in use. Its very early use by this nation, however, is only a supposition, and against the theory is

the fact that the custom did not extend to
neighboring nations as it did in other parts
of Europe soon after the introduction of
the weed from America.

In China the use of tobacco is common
to both sexes, to all the provinces, to the
diverse classes of society, and to nearly all
ages. Even young girls of eight and ten
years smoke long pipes. Two kinds of
the plant are cultivated in the country, the
nicotiana sinensis and the *nicotiana fruti-
cosa.* They grow in nearly all the provinces of the Celestial Empire, but the cultivation is generally made on a small scale.
Each family grows in the garden that sur-
rounds its house the plants necessary for
its yearly consumption. However, three
provinces are particularly favorable for
the production, and they are about the
only ones that furnish the three or four
preferred brands which are sold in the
various markets. These provinces are
Che-kiang, Hoo-pe, and Quang-tong, where
the tobacco is colored in four different
shades, yellow, violet, black, and red.
At Canton about ten qualities are sold, but
only four of them are generally used.
Among these tobaccos several have been
dipped in a solution of opium. This
plunging gives them a more reddish color
and a slight opium taste. As for the Japanese, they are great smokers and cultivate a
particular kind of tobacco, the *nicotiana sinensis,* the leaves of which they cut into

exceedingly thin fibres. This tobacco, which is yellow and as fine as hair, is mild and of very agreeable flavor. The use of tobacco began at about the same epoch in China and Japan—that is, toward the year 1574. The Arabs at Cairo smoke the best quality of tobacco; sometimes they perfume it with rose water and mix amber-scented pastilles with it in their chibouks. The smoke that they thus inhale is impregnated with agreeable odors.

The use of tobacco among the American Indians was prevalent from ancient antiquity, the custom being to inhale it through the nostrils by means of a small hollow-forked cane, shaped like a pitchfork, the single fork being placed in the fire and the shorter tubes up their nostrils. This instrument was called *tobago*, and from this term comes the word tobacco applied to the weed itself. At the time when Columbus discovered the New World the habit of smoking was common in South America. An historian, writing of these times, says of the tobacco plant, "It is called *petun* by the Brazilians; *tapaco* by the Spaniards; the leaves of which well dried they place in the open (widespread) part of a pipe, of which (being burned) the smoke is inhaled into the mouth by the more narrow part of the pipe, and so strongly that it flows out of the mouth and nostrils, and by that means drives out humors." When Cortez made

the conquest of Mexico, in 1519, smoking
was an established custom among the people ; Montezuma would have his pipe
brought to him with much ceremony by
leading ladies of his court, indulging in
the luxury after dinner and washing out
his mouth with scented water.

The early histories of the New World are
full of curious and interesting allusions to
the smoking habit. One of the members
of the expedition of 1584, under the aus-
pices of Sir Walter Raleigh, states that the
Indians looked upon tobacco as a gift from
the Great Spirit for their especial enjoy-
ment. They burned it as a sacrifice, threw
it into the air and water to quell a storm at
sea, after an escape from danger they also
threw some into the air. "We ourselves,"
says the first writer about Virginia, "during
the time we were there used to suck it after
their manner, as also since our returne, and
have found many rare and wonderful ex-
periments of the vertues thereof ; of which
the relation would require a volume by
itselfe ; the use of it by so manie of late, men
and women, of great calling as else, and
some learned phisitions also, is sufficient
witnes." Another author, after stating
that the "salvages" call tobacco *apooke*,
says : "The salvages here dry the leaves
of this *apooke* over the fier, and sometimes
in the sun, and crumble yt into poudre,
stalks, leaves, and all, taking the same in

pipes of earth, which very ingeniously they
can make."

The use of tobacco among the Indians
has been connected with their religious
worship. No treaty can be ratified without smoking the pipe of peace. Wilson, in
his Prehistoric man, says : " In the belief
of the ancient worshipper, the Great Spirit
smelled a sweet savour as the smoke of the
sacred plant ascended to the heavens ; and
the homely implement of modern luxury
was in their hands a sacred censor, from
which the hallowed vapour rose with as fit-
ting propitiatory odours as that which perfumes the awful precincts of the cathedral
altar, amid the mysteries of the church's
high and holy days." The Indian calumet,
or pipe of peace, is a sacred pipe ornamented
with the war eagle's quills. It is never allowed to be used on any other occasion than
that of peace-making. When a treaty is
made the chief brings it out, unfolds the
many bandages which are carefully kept
round it, and the pipe is passed to the different chiefs, each one in turn taking only
one breath of smoke through it.

" From the red stone of the quarry,
With his hand he broke a fragment,
Moulded it into a pipe-head,
Shaped and fashioned it with figures.
From the margin of the river
Took a long reed for a pipe-stem,
With its dark, green leaves upon it ;
Filled the pipe with bark of willow,
With the bark of the red willow ;

> Breathed upon the neighboring forest,
> Made its great boughs chafe together,
> Till in flame they burst, and kindled ;
> And erect, upon the mountains,
> Gitche Manito the Mighty
> Smoked the calumet, the Peace pipe,
> As a signal to the nations."

The seeds of the tobacco plant were first brought to Europe by Gonzalo Hernandez de Oviedo, who introduced it into Spain about 1560, where it was first cultivated as an ornamental plant, until another Spaniard claimed that it possessed medicinal virtues. It was introduced into Italy in 1560.

It has generally been claimed that Sir Walter Raleigh was the originator of smoking in England, and he certainly made the custom fashionable, but tobacco was really introduced into England by Mr. Ralph Lane, who was sent out by Raleigh as Governor of Virginia, and who returned to England in 1586. When Raleigh's servant for the first time saw his master smoke, he drenched him with beer, thinking he had got afire !

The tobacco plant was known in England before this date ; one writer claims it came into England in 1577, Taylor, the Water-poet, says 1565, and a Dutch author says 1576. To Raleigh, however, must be given the credit for introducing the habit of smoking.

> " While yet the world was young, the gods on high
> Bestowed the gift of wine upon the earth ;

The wide world rang with jocund minstrelsy,
  And Laughter shouting ushered in the birth.
The boon was suited to the youth of man ;
  But as the weight of years upon him grew,
And wise and sorrowful old age began,
  The gods consulted and devised anew.
Then was revealed the solace that should bless ;
  Gray-bearded Contemplation wore a smile ;
Grief raised her hands in trembling thankfulness :
  And all confessed they now might bear awhile.
Said Zeus : 'Love, Wine and Learning, 'tis, but three;
The race is dying—let Tobacco be.'"

As already stated, when the tobacco plant
first became known it was supposed to
possess almost miraculous healing powers,
and was designated herba panacea, herba
santa, sana sancta Indorum.

The early Spanish physicians adminis-
tered it to sick people by inhalation ; when
the patient was thoroughly intoxicated by
it, a cure was generally affected.

Spenser called it "divine tobacco" and
William Lilly called it "our holy herb
nicotian." Their references show that the
plant was used for wounds. Spenser says
it was brought to heal Timais, "who all
this while lay bleeding out his heart-blood
neare." And Lilly says it was used to heal
a lover whose hand had been wounded
with a spear. "Robinson Crusoe" speaks
of the Brazilians taking no physic, but using
tobacco for almost all their distempers.
Suffering from some ailment, he says he
took a roll of tobacco from one of his
chests. "I first took a piece of a leaf, and
chewed it in my mouth, which, indeed, at

first almost stupefied my brain, the tobacco
being green and strong, and I had not been
much used to it; then I took some and
steeped it an hour or two in some rum, and
resolved to take a dose of it when I lay
down; and lastly, I burnt some upon a pan
of coals, and held my nose close over the
smoke of it as long as I could bear it, as
well for the heat as the virtue of it, and I
held out almost to suffocation."

He took a dose of the rum and tobacco,
fell into a sound sleep and when he awoke
he found himself "exceedingly refreshed"
and his spirits "lively and cheerful."

Whilst the great plague raged in London,
tobacco was recommended by the faculty,
and generally taken as a preventive against
infection. It was popularly reported that
no tobacconists or their households were
afflicted by the plague. Physicians who
visited the sick took it very freely; the men
who went round with the dead carts had
their pipes continually alight. This gave
tobacco a new popularity, and it became
a popular cure of the day. The school boys
at Eton were obliged to smoke in the school
every morning, and they were whipped if
they did not.

A doctor, in the time of James I.,
directed a patient who was suffering from
an inflamed tooth to smoke without inter-
mission until he had consumed an ounce
of tobacco. The man was accustomed to
smoke, and therefore took twenty-five pipes

at a sitting, which had ⸱
him.

At the present day phy⸱
ing tobacco in conjunction
for poultices, the weed havı ɡ a vᴏ⸱ɥ ᴏᴏᴏ⸱⸱
ing effect on inflammation, as the writer of
this book can testify from his own experi-
ence. Several cases are reported of persons
being cured of poisoning by arsenic after
having swallowed a quantity of tobacco
juice.

The Indians of the forests oɪ the Orino-
co smoke tobacco, not only to produce an
afternoon nap, but also to induce a state of
quiescence, which they call dreaming with
the eyes open.

Commodore Wilkes, in his "Narrative of the United States Exploring
Expedition," tells a tough yarn of how the
tobacco habit once saved a man's life. He
says a Feejee islander told him how he and
his fellow-cannibals had captured a crew
from a ship that had been driven on the
coast. " What did you do with the men ?"
the Commodore inquired. " We eat 'em—
they good," said the Feejee, grinning.
The Commodore felt a qualm, as he inquired, faintly, " Did you eat them all ?"
" Yes, we eat all but one." "And why did
you spare that one ?" asked Wilkes. " Because he taste too much like tobacco ;
couldn't eat him nohow."

In the early days there were various
severe edicts issued against the use of

,pes, Urban VIII. and
.lminated against it the
Church ; the priests and
key declared smoking a
crime, one sultan decreeing its punish-
ment by the most cruel kinds of death ;
the pipes of smokers were thrust through
their noses in Turkey, and in Russia,
in the seventeenth century, the noses of
smokers were cut off. In the Swiss
canton of Berne smoking ranked in the
table of offences next to adultery. James
I. of England issued his famous counterblast against the weed. He called it the
" lively image and pattern of hell, because it
was a smoke." His unwarrantable perse-
cution of the tobacco plant led him to raise
the importation duty from twopence per
pound to the monstrous sum of 6s. 10d.

Smoking had become so common in England in 1621 that one member of the House
of Commons wanted tobacco banished out
of the kingdom. He had been very much
shocked to see ploughmen smoke as they
were at their plough! Smoking was for-
merly forbidden among schoolmasters. In
the rules of a school at Chigwell, founded
in 1629, it was declared that the master
" must be a man of sound religion, neither
Papist nor Puritan, of a grave behavior,
and sober and honest conversation, no tip-
pler or haunter of ale-houses, and no puffer
of tobacco."

It was not at first allowed to be smoked

in ale-houses. Says an
an innkeeper : " You .
willingly suffer to be u⁞
taken, any tobacco wit⁞
cellar, or other place, there⁞ ⁞⁞⁞ ⁞⁞⁞⁞⁞⁞⁞⁞."
But the people knew what they wanted,
and puffed smoke into the faces of their
critics. Many a good sermon was preached
by the weed. Here is one :

> " This Indian weed, now withered quite,
> Though green at noon, cut down at night,
>> Shews thy decay,
>> All flesh is hay,
> Thus think, and smoke tobacco.
>
> " The pipe so lily white and weak,
> Doth thus thy mortal state bespeak,
>> Thou art e'en such,
>> Gone with a touch !
> Thus think, and smoke tobacco.
>
> " And when the smoke ascends on high,
> Then dost thou see the vanity
>> Of worldly stuff
>> Gone with a puff !
> Thus think, and smoke tobacco.
>
> " And when the pipe grows foul within,
> Think of thy soul begrimed with sin,
>> For then the fire
>> It does require !
> Thus think, and smoke tobacco.
>
> " And seest thou the ashes cast away,
> Then to thyself thou mayest say
>> That to the dust
>> Return thou must!
> Thus think, and smoke tobacco."

I.

IE HOTTENTOTS, THE
FRICA AND CURIOUS
OVER THE WORLD—A
GREENLANDER ENJOYS A SWALLOW OF
NICOTINE—MORAL PHILOSOPHERS, FA-
MOUS POETS, AND GREAT LEADERS OF
MEN HAVE BEEN LOVERS OF THE
WEED—A CHRONOLOGY OF TOBACCO
FROM ITS DISCOVERY TO THE PRESENT
TIME—HOW SMOKE COULD BE MADE
TO AFFECT LEGISLATION—FAMOUS
FRENCH SMOKERS—SMOKING AMONG
WOMEN.

Toward the end of the sixteenth century
tobacco penetrated by the Bosphorus to
Ispahan and even to India. No one knows
who first carried it to the true believers,
but it is certain that the Mussulmans found
in its fervent practice a foretaste of the
delights promised by the prophet to every
good servitor of Islam, and that tobacco
mingled its aromas in all the Oriental
pleasures. But the Mohammedans were
more severe than the Christians against
the crime of smoking. Amurat IV. even
condemned smokers to death. In Persia
the most rigorous penalties were decreed
against all those who did not abominate
the forbidden plant. All these persecutions
did not prevent the Orientals from becoming as enthusiastic over the weed as were
the Europeans. In fact, the passion of the

Orientals for tobacco is beyond all expression. Its use, like that of perfumes, is universal. In all classes men, women and children smoke without distinction, not only after meals, but at all hours. The poorest person in that happy country always finds tobacco enough to fill his pipe and indulge in a *kief*. Without being lazy, the Turk does not like to hurry. "Haste comes from the devil and patience from God," he says.

In Persia tobacco grows easily and almost everywhere. It is a real dead leaf when it is dried, and is not so strong as ours. But the Persians prefer it so that they can smoke it all day long. For this purpose they crumble the tobacco very fine and wet it a little in order not to have it burn too quickly. The excessive use of this plant dries up the Persians and weakens them. They admit this fact, but when asked why they do not quit the habit, reply: "There is no joy for the heart except by tobacco."

There are three principal kinds of tobacco in the Orient, but the quality is variable. At Constantinople the dealers sell smoking tobacco cut very fine in long, silky bunches of blond color. *Iavach* is the name for the mild quality; *orta* designates the average strength; *tokan aklen* the sharp tasting, and *sert* the very strong. The Levant tobacco is the most sought after, it being very mild. It is of a yellowish green color, breaks easily, and burns to the end with white ashes. There is also a Levant tobacco called *karasson*, or black tobacco, which is perfumed, but very strong. The mildest kind is the *sultanieh.* The *tombaki*, a tobacco cultivated in the south of Persia, and especially at Schiraz, is rather a sort of illusive preparation than tobacco properly called. It is composed of tobacco, pieces of sandalwood or aloes wood, rose leaves, haschisch, and opium. In the bazaar at Teheran, the alley where the tobacco dealers congregate is impregnated with the strong aroma that escapes from the goatskin sacks containing the tombaki, so strong that it can be inhaled only after being softened in water mixed with essence of roses.

Tobacco is to-day one of the principal products of Hindoostan, where it was introduced by the Europeans somewhere between 1555 and 1627. A great quantity is consumed in India, and very good quality is grown in several provinces, particularly in Guzerat, where the *zerd*, as it is called on account of its yellow color, is highly appreciated. The leaf is small, and has a balsamic perfume and a gentle taste, while its smoke spreads an agreeable violet perfume. The Hindoos mix sugar, nutmegs, and bananas with their tobacco ; they pound these materials in a mortar and add thereto rose water. This is the tobacco that they smoke in their pipes. A very high quality of the plant is also cultivated in Malwah.

The Negritos, in Luzon (one of the Philippines), scarcely ever stop smoking cigars, putting the lighted end in their mouths. The Hottentots barter their wives for tobacco, and when they cannot obtain it, fill their pipes with a substitute of dried dirt. In the snowy regions of the Himalaya, tiny smoking tunnels are made in the frozen snow, at one end of which is placed some tobacco, along with a piece of burning charcoal, while to the other end the mountaineers place their mouths, and lying on their stomachs, inhale the smoke of the glowing weed. The Patagonian lights a pipe, throws himself down with his face toward the ground, and swallows several mouthfuls of smoke in a manner which produces a kind of intoxication lasting for several minutes. The inhabitants of the Cook peninsula, in Australia, are passionate smokers. Their pipe—a bamboo three and a half feet long and four inches in diameter—passes round the company after one of the persons present has filled it with smoke from a tube. The Wadschidschi, dwelling by the banks of the Tanganyika Lake, neither chew nor snuff nor smoke their tobacco, but carrying it in a small vessel, the savage pours water upon it, and presses out the juice, with which he contrives to fill both nostrils, keeping it there by means of wooden pegs. The Kaffirs,

who cannot get snuff as fine and as pungent
as they wish, rub the already prepared
mass between stones, and mix it with a
kind of pepper and some ashes. The
blacks in Dschesire mix their tobacco with
water and natron, so as to form a kind of
pap which they call bucka. They take a
mouthful and roll it about for a time with
their tongue. There are regular bucka
parties given. In Paraguay it is chiefly
the women who chew ; and travellers have
often described their emotions when on entering a house, a lady dressed in satin, and
adorned with precious stones, comes toward
them, and, before holding out her mouth
to be kissed, as the usual welcome, pulls
the beloved tobacco quid from her cheek
pouch. Some South American tribes eat
the tobacco cut into small pieces. Finally,
there is a traveller's story told of certain
Esquimaux tribes that, if true, is not a
little remarkable. When a stranger arrives
in Greenland, it is said that he finds himself immediately surrounded by a multitude
of natives, who ask his permission to drink
the oil which remains in the stem of his
pipe. And it is stated that the Greenland-
ers smoke for no other purpose than to enjoy afterward the swallowing of that acrid
and poisonous matter which is so disagreeable to us.

In Nicaragua, the dress of the urchins,
from twelve or fourteen downward, consists generally of a straw hat and a cigar,

a costume which is airy, picturesque and cheap.

The following table will show, at a glance, some of the most remarkable events in the history of the weed ·

CHRONOLOGY OF TOBACCO.

A.D.

1496. Romanus Paine published the first account of tobacco, under the name *Cohoba.*

1519. Tobacco discovered by the Spaniards near Tabasco.

1535. Negroes cultivated it on the plantations of their masters.

1535. It was used at this time in Canada.

1559. Tobacco introduced into Europe.

1570. Tobacco smoked in Holland out of tubes of palm-leaves.

1585. Clay pipes noticed by the English in Virginia.

1585. First clay pipes made in Europe.

1599. Tobacco prohibited in Persian Empire.

1601. Tobacco introduced into Java. Smoking commenced in Egypt about this time.

1604. James I. laid heavy imports on tobacco.

1615. Tobacco first grown in Holland.

1616. The colonists cultivated tobacco in Virginia.

1619. James I. wrote his " Counterblast."
1620. Ninety young women sent from England to America, and sold to the planters for tobacco at 120 pounds each.
1624. The Pope excommunicated all who should take snuff in church.
1634. A tribunal formed at Moscow to punish smoking.
1653. Smoking commenced in Switzerland.
1669. Adultery and fornication punished in Virginia by a fine of 500 to 1000 pounds of tobacco.
1689. Tubes containing pieces of sponge invented for smoking tobacco.
1691. Pope Innocent XII. excommunicated all who used tobacco in St. Peter's Church at Rome.
1724. Pope Benedict XIV. revoked Pope Innocent's bull of excommunication.
1732. Tobacco made a legal tender in Maryland at one penny per pound.
1789. King of France derived an income of $7,500,000 from tobacco.
1789. Exports of tobacco from the United States, 90,000,000 pounds.
1828. Tobacco revenue in the State of Maryland, $27,000.
1830. Revenue from tobacco and snuff in Great Britain, $12,000,000.
1834. Value of tobacco used in the United States estimated at $15,000,000.
1889. Value of tobacco product in the United States, $43,666,665.

Among the famous men who have drawn
inspiration and consolation from the pipe
were Milton, who had his pipe and a glass
of water just before he retired for the night.
Philosophers have drawn their best similes
from their pipes. How could they have
done so, had their pipes first been drawn
from them ? We see the smoke go upward
—we think of life; we see the smoke-
wreath fade away—we remember the morning cloud. Our pipe breaks—we mourn
the fragility of earthly pleasures. We
smoke it to an end, and tapping out the
ashes remember that "Dust we are, and
unto dust we shall return." If we are in
love, we garnish a whole sonnet with
images drawn from smoking, and first fill
our pipe, and then tune it. That spark
kindles like her eye, is ruddy as her lips;
this slender clay, as white as her hand, and
slim as her waist; till her raven hair grows
gray as these ashes, I will love her. This
perfume is not sweeter than her breath,
though sweeter than all else. The odor
ascends into the brain, fills it full of all
fiery, delectable shapes, which delivered
over to the tongue become delectable wit.

Paley, the moral philosopher, was an excellent companion. On a cold winter's
night he would stir the fire and fill a long
Dutch pipe. He formally declined any
punch, but nevertheless drank it up as fast
as his glass was replenished ; in fact, he
would smoke any given quantity of tobac-

co, and drink any given quantity of punch.
Dr. Parr's particular fondness for smoking
was so well known that wherever he dined,
he was always indulged with a pipe. Even
George IV. provided him with a smoking-
room, saying, "I don't like to be smoked
myself, doctor, but I am anxious that your
pipe shall not be put out." When a certain lady absolutely refused him permission
to smoke in her parlors he was very wroth,
but contented himself with calling her
"the greatest tobacco-stopper in all England."

Blücher, the famous military leader, had
a servant or pipe-master named Hennemann. At the battle of Waterloo the pipe-
master had just handed a pipe to his master
when a cannon-ball caused Blücher's horse
to spring aside and the pipe was broken
before the old hero had a chance to take a
single puff. "Fill another pipe for me,"
said Blücher, "and keep it lighted until I
come back in a moment, after driving away
the French rascals." The chase lasted not
only a moment, but a whole hot day ; then
Blücher met Wellington, who asked him
about his previous position. Blücher went
to the spot where he had halted in the
morning. There stood a man with his head
bound up and his arm wrapped in a hand-
kerchief. He was smoking a long and
dazzlingly white clay pipe. "Good God,"
exclaimed Blücher, "that is my servant,
Christian Hennemann. What a strange

look you have, man ! What are you doing
here?" "Have you come at last?" answered
Christian, in a grumbling tone; "here I
have stood the whole day, waiting for you.
One pipe after another has been shot away
from my mouth by the accursed French.
Once even a blue bean [bullet] made sad
work with my head, and my fist has got a
deuce of a smashing. That is the last whole
pipe, and it is a good thing that the firing
has stopped ; otherwise the French would
have knocked this pipe to pieces, and
you must have stood here with a dry
mouth."

An art critic has observed that the difference between two great French painters,
Decamps and Horace Vernet, was due to
their habits as users of tobacco. The French
Murillo, the wonderful colorist, Decamps,
smoked a pipe ; Vernet toyed with the
cigarette.

A lady expressed surprise at seeing Gui-
zot smoking. "What !" she exclaimed,
"you smoke, and yet have arrived at so
great an age ?" "Ah ! madame," replied the
venerable statesman, "if I had not smoked,
I should have been dead ten years ago."

Victor Hugo was a veteran smoker.
Buckle, the great historian of civilization,
found it so imperious a necessity to have
his three cigars every day, that he said he
could neither read, write, nor talk if com-
pelled to forego, or even to miss the usual
hour for indulging in them. A traveller,

who once met him in the East, found him smoking Latakia out of a large red-clay pipe with an extremely long cherry stalk, and drinking coffee, *a la turque*, with evident satisfaction.

It is significantly stated that Richard Fletcher, a courtly Bishop of London (1596), by the use of tobacco, "smothered the cares he took by means of his unlucky marriage."

Raleigh smoked in his dungeon in the Tower, while the headsman was grinding his axe. Cromwell loved his pipe and dictated his despatches to Milton over sweet-smelling nicotine. Robert Hall smoked in his vestry.

Carlyle has said that tobacco smoke is good, because it allows men to sit silent together without embarrassment. When a man has said what he has got to say he can hold his peace and take to his pipe. He says that such a practice could be wisely introduced into Parliaments, where there should be a minimum of speech and the soothing and clarifying influence of tobacco smoke.

This idea seems to have been previously acted upon by Frederick William I., King of Prussia, who founded the Tabaks-Collegium, which was a sort of smoking Parliament, where grave political discussions were carried on by the members as they puffed at their clay pipes. This smoking room was supplied with plenty of pipes and tobacco, and refreshments, consisting of beer, cold bread and beef.

If a certain anecdote is true, tobacco seems to have had its influence on the speech of the great Carlyle himself. There is a legend to the effect that on the one evening passed at Craigenputtock by Emerson, in 1833, Carlyle gave him a pipe, and, taking one himself, the two sat silent till midnight, and then parted, shaking hands, with congratulations on the profitable and pleasant evening they had enjoyed.

Tennyson, the Poet Laureate, is a great smoker, and does not affect Havana in any of its various forms. His joy is in a pipe of genuine Virginia tobacco. He prefers a pipe, the common clay pipe being his choice. He has a great many kinds of pipes, mostly presents from admirers and friends. When smoking with his friends, in his den, which is at the top of the house, he sits with a box full of clay pipes at his feet. Filling one of these, he smokes until it is empty, breaks it in twain, and throws the fragments into another box prepared for their reception. Then he pulls another pipe from its straw or wooden enclosure, fills it, lights it, and destroys it as before. He will not smoke a pipe a second time. James Payn smokes constantly, using a pipe and Latakia tobacco. The doctors once told him that the use of such tobacco would kill the strongest man in the world, but he has been smoking it for a quarter of a century with impunity.

Among French authors Zola has said that he does not believe the intelligence and creative strength of man are injured by smoking. François Coppée smokes cigarettes all day, but throws each one away after a few puffs. H. Taine smokes cigarettes, considering the habit a pastime in moments of thoughtlessness and intellectual waiting. Andrew Theniret once said that he was not a member of the French anti-tobacco league because he was passionately fond of smoking. This was in answer to a question from the league concerning the effects of smoking on the mind and body. "Two years ago," he wrote in reply, "your president asked me to write a story about the acute sufferings of the young smoker. I did it. After publishing the story I received a silver medal from your league. That is all I ever had to do with the enemies of tobacco."

The daughters of Louis XIV. of France imitated some English ladies of the period, and indulged in a pipe. They were in the habit of indulging in a sort of orgie in their own apartments after supper, and one evening were found in the act of drinking brandy and smoking pipes, which they had borrowed from the officers of the Swiss Guard. The present Empress of Austria, having little taste for reading, loves, when at home, to loll back in an easy-chair or lie on a sofa and puff cigarettes. She hates brilliant assemblies, and loves to talk with a congenial companion on equestrian subjects, being very fond of horses.

About the year 1700 "Tom Brown," an anonymous wit of the day, wrote this peculiar letter to an imaginary ancient dame who smoked tobacco: "Though the ill-natured world censures you for smoking, yet would I advise you, madam, not to part with so innocent a diversion. In the first place, it is healthful; and, as Galen rightly observes, is a sovereign remedy for the toothache, the constant persecutor of old ladies. Secondly, tobacco, though it be a heathenish word, is a great help to Christian meditations, which is the reason, I suppose, that recommends it your parsons, the generality of whom can no more write a sermon without a pipe in their mouths, than a Concordance in their hands; besides, every pipe you break may serve to put you in mind upon what slender accidents man's life depends. I knew a dissenting minister who, on fast days, used to mortify upon a rump of beef, because it put him, as he said, in mind that all flesh was grass; but I am sure much more is to be learnt from tobacco. It may instruct you that riches, beauty, and all the glories of the world, vanish like a vapor. Thirdly, it is a pretty plaything. Fourthly, and lastly, it is fashionable—at least, 'tis in a fair way of becoming so."

When snuff came into use it found favor with the fair sex. One English dame had for her maxim:

" She that with pure tobacco will not prime
Her nose, can be no lady of the time."

In Havana a small and very fine kind of cigars are made for the use of ladies, and are called " Queens." Women there smoke as freely as men, and in a full railroad car, every person, man, woman, and child, may be seen smoking. To put up a sign, " No smoking," and enforce the rule would ruin the road.

At Manilla about 12,000 women are employed in the cigar manufactories. Paper cigarettes are chiefly smoked by the men; the women prefer the largest cigars they can get. The women of Johore are often seen seated together weaving mats, and each with a cigar in her mouth.

The wife of General Jackson, seventh President of the United States, was an exemplary woman in all the relations of life, but, in the homely fashion of the time, she used to join her husband and guests in smoking a pipe after dinner and in the evening.

The enemies of the weed say that tobacco is a poison because animals will not use it. A Berlin professor, an artist, however, who has lately experimented in the Zoological Gardens, declares that common brown bears are genuine enthusiasts for tobacco.

"When I puff my cigar smoke into their cage," he remarks, "they rush to the front, rubbing their noses and backs against the bars through which the smoke has penetrated." The professor, with some temerity, once experimented on the lion. The creature was asleep, and this was the moment selected for puffing a volume of tobacco smoke in his face. Did he at once wake up with a savage growl, lash his tail, and, springing at the bars, shake the massive iron? Not at all. He awoke and "stood on his legs," which seems a natural enough attitude to adopt, and "sneezed powerfully." Then he quietly laid down on his side and "elevated his nose, as if asking for a second dose." It may be news to some naturalists to hear that goats, stags, and llamas all devour tobacco and cigars with remarkable satisfaction. It is certainly somewhat of a waste of the material to let a prime Havana be "bolted" in one gulp by an antelope; but the professor was actuated by a praiseworthy desire to discover scientific facts, and also by a wish to get on good terms with creatures whom it was his business to sketch. "I made a personal friend," he writes, "of an exceedingly malicious guanaco, or wild llama, by simply feeding him again and again with tobacco.

"Brother Gray," said one clergyman to another, "is it possible you smoke tobacco? Pray give up the unseemly practice. It is

alike unclerical and uncleanly. Tobacco !
Why, my dear brother, even a pig would
not smoke so vile a weed." Brother Gray
delivered a mild outpouring of tobacco
fumes, and then as mildly said, " I suppose,
Brother Curtis, you don't smoke ?" " No,
indeed," exclaimed his friend, with virtu-
ous horror. Another puff or two, and then
Brother Gray, who prefers the Socratic
method of argument, rejoined, " Then,
dear brother, which is more like the pig,
you or I ?"

III.

THE "FAIRY PIPES" OF IRELAND — QUEER
PIPES USED BY THE RUSSIANS, GERMANS,
FRENCH, CHINESE, JAPANESE, NATIVES OF
SOUTH AFRICA, ETC. —HOW ENGLISH CHIL-
DREN SMOKED ON THEIR WAY TO SCHOOL—
"KEMBLE PIPES"—MUSICAL PIPES—THE
PIPE-SMOKER IS A PHILOSOPHER—FAMOUS
PIPE-SMOKERS.

Of the different kinds of pipes there
were the "fairy pipes" of Ireland; they
were small and did not hold quite as
much tobacco as our modern meerschaums.
The Irish peasantry believed them to have
been of fairy-demon origin; when found
they would be at once broken as a
kind of retort to some mischievous
trick which their supposed owners had
played. Pipes were first regularly manufactured in England in 1619. The pipe
became an object of much inventive inge-
nuity, and it varied as greatly in material
as in form—wood, horn, bone, ivory, precious stone, valuable metals, amber, glass,
porcelain, and, above all, clay being the
materials employed in various forms. By
degrees, pipes of special form and material
came to be associated with particular
people, so that now we have the elongated
painted porcelain bowls and pendulous
stem of the German peasantry, the red
clay bowl and long cherry-wood stem of

the Turk, and the very small metallic bowl
and cane stem of the Japanese.

The Dutch borrowed the art of pipe-
making from the English in 1748. A
Dutchman visiting England in the time of
Queen Elizabeth, was surprised to see the
English " draw the smoke into their mouths
(through pipes made of clay), which
they puff out again through their nostrils,
like funnels." During the reign of the
Great Plague (1644 and 1666) there was a
great deal of smoking to ward off the dis-
ease, and a large number of pipes have
been discovered in and about London belonging to this era. Some of the early
pipes were made of silver, but the ordinary
sort were made from a walnut shell and a
straw. Clay pipes soon became cheap and
common and were passed from man to
man round the table. The Dutch loved
their mahogany pipes so much that they
carried them in ornamented wooden cases,
which were sometimes inlaid with brass,
on which was engraved some proverb or
scriptural motto. On the pipe-case of the
famous Admiral Van Tromp was this in
scription: " When a man has the right way
taken, death has no fears for him."

The best French pipes are made of porcelain, and some are adorned with enamelled portraits and beautiful heads. Others
are made of various kinds of earth, or
earthy compounds, compressed in moulds
by the potter, and afterwards cut in deeper

relief by hand. Some are made of rare kinds of wood and lined with clay, and others are fashioned in elegant shapes from masses of agate, amber, crystal, carnelian and ivory, as well as the various kinds of pure or mixed metals. The handsomest pipes of French manufacture are bought by foreigners, most Frenchmen contenting themselves with the ordinary pipe of soft porous clay. The pipe commonly in use among the Russians is made of wood, tipped with red copper, and lined with a thin sheet of tin, rudely nicked and turned over at the rim. The stem is of dogwood, and is tied to the pipe by a rough thong of leather, to which is affixed a pick, made of copper wire, to clear out the pipe when necessary.

The Germans long used a beautiful pipe, carved by the herdsmen and peasants of the Black Forest from the close-grained and gnarled root of the dwarf-oak. This wood is hard enough to resist the action of fire, becoming but slightly charred by years of use. The carvings represented boar hunts, encounters with wolves, jowling, and the exploits of robbers. The ordinary German pipe of porcelain consists of a double bowl, the upper one containing the tobacco, which fits into a spout or socket, and allows the oil to drain into the lower bowl, which is generally held in the hand of the smoker; the tube of wood, usually formed of cherry-tree, is easily moved, by which it may be cleaned.

The pipe of the Egyptians is usually between four and five feet long : some pipes are shorter, and some are of greater length. The common kind are made of wood, and the greater part of the stick, from the mouthpiece to about three-quarters of its length, is covered with silk, confined at each end by gold thread, or by a tube of gilt silver. The covering was originally designed to be moistened with water, in order to cool the pipe, and consequently the smoke by evaporation. In smoking, the people of Egypt and other countries of the East, draw in their breath freely, so that much of the smoke descends into the lungs. The terms which they use to express " smoking tobacco" signify " drinking smoke," or " drinking tobacco."

The natives of South Africa ceremoniously smoke a *daghapipe*, made out of bullock's horn, and use a species of hemp instead of tobacco. Each individual receives the pipe in turn, opens his jaws to their full extent, and placing his lips to the wide mouth of the horn, takes a few pulls and passes it on.

The pipe used by the Chinese has a straight stem from three to five feet in length. To the stem of the pipe is sometimes attached tassels and silken pendent ornaments. The stem is usually made of bamboo. Both men and women smoke, and pipe-sellers walk through the streets, plying their trade. One kind of Chinese pipe is made of brass and constructed on the principle of the *hookah*, with a large trumpet-shaped receptacle filled with water, and a cup for tobacco. The pipe is provided with a base to stand upon the table, and the smoke is drawn through water. Only a few whiffs are taken at a time, the tobacco used being cut into very fine shreds. Japanese pipes are often made of silver, inlaid with flowers and insects in enamelled copper. The central portion is formed of cane, for convenience of holding.

The most luxurious and elaborate form of pipe is the Persian kalydn, hookah, or water-tobacco pipe. This consists of three pieces, the head or bowl, the water-bottle or base, and the snake, or long flexible tube ending in the mouth. The tobacco, which must be previously prepared by steeping in water, is placed in the head and lighted with live charcoal, a wooden stem passes from the bottom down into the water which fills the base, and the tube is fitted to a stem which ends in the bottle above the water. Thus the smoke is cooled and washed before it reaches the smoker by passing through the water in the bottle and by being drawn through the coil of tube, frequently some yards in length. The bottles are, in many cases, made of carved and otherwise ornamented cocoa-nut shells, whence the apparatus

is called nargila, from nargil, a cocoa-nut.
Silver, stone, damascened steel and precious stones are freely used in the making
and decoration of these pipes for wealthy
smokers.

We are so accustomed to hearing about
the ill effects of smoking on the young
that it seems strange to read that in England, in the seventeenth century, children
going to school carried with their books a
pipe of tobacco, which their mothers took
care to fill early in the morning. At a
certain time in school every one laid aside
his book to light his pipe, the master
smoking with them, and teaching them
how to hold their pipes and draw in the
tobacco, thus getting them used to the
weed from their youth as a practice absolutely necessary for a man's health.

What was once known as the Kemble
pipe has a curious history. Those pipes derived their name from a poor Roman Catholic priest who was executed in 1679, he
having been implicated in the plot of Titus
Oates. While marching to the scaffold he
smoked a pipe of tobacco. In memory of
this, the people of Herefordshire to this
day call the last pipe they take at a sitting,
a Kemble pipe.

George Augustus Sala, some years ago in
"Household Words," lamented the disappearance of the old church-warden pipe, or
"yard of clay." He said there were a host
of inventions for emitting the fumes of to-

bacco and that English gentlemen had got
in the habit of smoking "black abomina-
tions, like Irish apple-women."

> " Little tube of mighty power,
> Charmer of an idle hour,
> Object of my warm desire,
> Lip of wax and eye of fire :
> And thy snowy taper waist,
> With my finger gently braced ;
> And thy pretty swelling crest,
> With my little stopper pressed ;
> And the sweetest bliss of blisses,
> Breathing from thy balmy kisses,
> Happy thrice, and thrice again,
> Happiest he of happy men,
> Who, when again the night returns,
> When again the taper burns,
> When again the crickets gay
> (Little cricket, full of play),
> Can afford his tube to feed
> With the fragrant Indian weed ;
> Pleasure for a nose divine,
> Incense of the god of wine,
> Happy thrice and thrice again,
> Happiest he of happy men."

Tobacco-pipes have contributed to amuse
non-smokers by being subservient to in-
genious tricks. An English tavern-keeper
amused his company with whistling of
different tunes ; he took up a pair of clean
tobacco-pipes, and after having slid the
small ends of them over a table in a most
melodious trill, he fetched a tune out of
them, whistling to them at the same time
in concert. The virtuoso confessed ingen-
uously, that he broke such quantities of
pipes that he almost broke himself, before
he brought this piece of music to any tol-

erable perfection. Balancing tobacco-pipes
was a novel feat introduced for London's
amusement.   In 1743 a fire-eater, in one of
his advertisements, notes among his other
performances, that he " licks with his naked
tongue red-hot tobacco-pipes flaming with
brimstone."

There seems to be a close connection between pipe-smoking and the philosophical
habit.   Captain Marryat says in "Jacob
Faithful :" "It is no less strange than true
that we can puff away our cares with tobacco, when, without it, they remain an
oppressive burden to existence.   There is
no composing draught like the draught
through the tube of a pipe.   The savage
warriors of North America enjoyed the
blessing before we did ; and to the pipe
is to be ascribed the wisdom of their coun-
cils, and the laconic delivery of their sentiments."

And "Sam Slick, the Clock-maker,"
says : "The fact is, the moment a man
takes to the pipe he becomes a philosopher.
It's the poor man's friend ; it calms the
mind, soothes the temper, and makes a man
patient under difficulties.   It has made
more good men, good husbands, kind masters, indulgent fathers, than any other
blessed thing on this universal earth."

"Sweet smoking pipe ; bright glowing stove,
Companion still of my retreat,
Thou dost my gloomy thoughts remove,
And pinge my brain with gentle heat.

43

"Tobacco, charmer of my mind,
When, like the meteor's transient gleam,
Thy substance, gone to air, I find
I think, alas ! my life's the same.

" What else but lighted dust am I ?
Thou show'st me what my fate will be ;
And when thy sinking ashes die,
I learn that I must end like thee."

When Lord Brougham was in the zenith
of his fame he was fond of smoking. He
would smoke a pipe after his labors in the
court room, another one after speaking in
the House of Commons, and another before
going to bed. Lord Clarendon, England's
Foreign Secretary, always smoked when
attending to his official business, and the
Foreign Office, while he was there, was
always pervaded with a strong aroma
of cigars. His despatches were generally
written between midnight and daybreak,
and during this time a cigar or cigarette
scarcely ever left his lips. He never felt
at ease at a diplomatic conference until
cigars were introduced, and this remark is
attributed to him : " Diplomacy is entirely
a question of the weed. I can always
settle a quarrel if I know beforehand
whether the plenipotentiary smokes Cav-
endish, Latakia or Shag. Tobacco is the
key to diplomacy."

Sir Isaac Newton, the great natural
philosopher, was a prince among smokers.
Some modern reformers say that tobacco
injures the teeth. Newton exposed this

44

fallacy, for he lived to a good old age and never lost but a single tooth. It is recorded of him that on one occasion, in a fit of mental abstraction, he used the finger of the lady he was courting as a tobacco stopper, as he sat and smoked in silence beside her! Professor Huxley, the modern philosopher, hated tobacco when a young man, but is now a lover of the weed. He says that smoking in moderation is a comfortable and laudable practice, and is productive of good. "There is no more harm," he says, "in a pipe, than there is in a cup of tea. You may poison yourself by drinking too much green tea, and kill yourself by eating too many beefsteaks. For my own part, I consider that tobacco, in moderation, is a sweetener and equalizer of the temper."

Charles Lamb confessed that he had been "a fierce smoker of tobacco." When he decided to give up smoking he compared himself to "a volcano burned out and emitting only now and then a casual puff." He called tobacco his "loving foe," his "friendly traitress," the "great plant," and attributed to it his chronic indisposition, which Carlyle says was really caused by his "insuperable proclivity to gin."

One day Lamb was puffing away at the strongest and coarsest preparation of the weed in company with Dr. Parr, who could only smoke the finest sorts of tobacco.

Parr asked Lamb how he had acquired
such "prodigious power" as a smoker.
"I toiled after it," replied the humorist,
with his habitual stutter, "as some men
t—t—toil after virtue." He once expressed
a wish to John Forster that his last breath
might be drawn through a pipe and exhaled
in a pun. This reminds one of the French
artist, Gavarni, who on his death-bed is
reported to have said to a friend: "I leave
you my wife and my pipe; take care
of my pipe."

Charles Kingsley, when he was too ex-
cited to write any more on the book he had
in hand, would calm himself down with a
pipe. He always used a long and clean
"church-warden" pipe, and these pipes used
to be bought a barrelful at a time ; when
there was a vast accumulation of old pipes,
enough to fill the barrel, they were sent
back again to the kiln to be rebaked, and
returned fresh and new. This gave the
novelist a striking simile ; in "Alton
Locke" he puts these words into the mouth
of James Crossthwaite: "Katie here be-
lieves in purgatory, where souls are burned
clean again, like 'bacca pipe." Speaking
of tobacco, another character in "West-
ward Ho" says: "The Indians always
carry it with them on their war-parties ;
and no wonder, for when all things were
made, none was made better than this, to
be a lone man's companion, a bachelor's
friend, a hungry man's food, a sad man's

cordial, a wakeful man's sleep, and a
chilly man's fire, sir; while for stanching
of wounds, purging of rheum, and set-
tling of the stomach, there's no herb like
unto it under the canopy of heaven.

> "All dainty meats I do despise,
> Which feed men fat as swine;
> He is a frugal man indeed,
> That on a leaf can dine.
>
> "He needs no napkin for his hands
> His fingers' ends to wipe,
> That keeps his kitchen in a box,
> And roast meat in a pipe."

Thackeray always began writing with a
cigar in his mouth and was a real devotee
of tobacco. A lady relates how, when a
young man, he was in Paris studying to
be a painter, he would dash into the room
where she was sitting, and say, "Polly,
lend me a franc for cigars." When dictat-
ing he would often light a cigar, and after
pacing the room for a few minutes would
put the unsmoked remnant on the mantel-
piece, and resume his work with increased
cheerfulness, as if he had gathered fresh
inspiration from the gentle odors of "sublime tobacco." Dickens was a smoker, and
we catch a glimpse of him smoking a fare-
well cigar with Thackeray at Boulogne.
There they conversed about a certain titled
lady, a singular character, who had made
Dickens smoke with her some cigars made
of negro-head, powerful enough, accord-

ing to his account, to "quell an elephant in six whiffs."

Clergymen have always been noted for their love of the weed. Richard Fletcher, Bishop of London in the time of Elizabeth, was the first Episcopal smoker in England. He was banished to Chelsea for marrying a second time and, as Camden says, "smothered his cares by the immoderate use of tobacco." He died suddenly, in his easy-chair, while smoking his pipe. The famous Bishop Burnet always smoked while he was writing ; in order to perform both operations comfortably, he would have a hole through the broad brim of his large hat, and, putting the stem of his long pipe through it, puff and write, and write and puff, with learned gravity. Dean Aldrich, the Oxford professor, was such an inveterate smoker that a student once laid a wager that he would be found smoking at ten o'clock in the morning, an early hour for him. The student went to the Dean's study at the appointed hour and related the occasion of his visit, to which the Dean replied, in perfect good-humor : "You see you have lost your wager, for I'm not smoking, but filling my pipe."

He was quite musical and composed "Hark, the bonny Christ Church bells ;" also, "A Smoking Catch, to be sung by four men smoking their pipes, not more difficult to sing than diverting to hear."

Some years ago Mr. Spurgeon preached

a sermon from the text : " I cried with my whole heart ; hear me, O Lord ! I will keep Thy statutes. I cried unto Thee ; save me, and I shall keep Thy testimonies." He spoke of the necessity of giving up sin and, at the conclusion of the discourse requested Rev. Mr. Pentecost, of Boston, who was present, to give the personal application of the sermon. Mr. Pentecost among other things spoke about the great struggle it had cost him to give up the use of tobacco. He said : "I liked exceedingly the best cigar that could be bought, but I felt that the Lord required me to give up smoking. So I took my cigar-box before the Lord and cried to Him for help." This help, he intimated, had been given, and the habit was renounced. Mr. Spurgeon, who is very fond of smoking himself, instantly rose at the conclusion of Mr. Pentecost's address, and, with a somewhat playful smile, observed that some men could do to the glory of God what in other men would be sin : " Notwithstanding what Brother Pentecost has said, I intend to smoke a good cigar to the glory of God before I go to bed to-night. If anybody can show me in the Bible the command, ' Thou shalt not smoke,' I am ready to keep it ; but I haven't found it yet. Why, a man may think it is a sin to have his boots blacked. Well, then, let him give it up, and have them whitewashed. I am not ashamed of anything whatever that I do, and I don't feel that smoking makes me ashamed, and therefore I mean to smoke to the glory of God." This manly utterance created considerable excitement in church circles, and Mr. Spurgeon wrote a letter to the " Daily Telegraph," in which he maintained his right to smoke. He said : " I will not own to sin when I am not conscious of it. There is growing up in society a Pharisaic system, which adds to the commands of God the precepts of men : to that system I will not yield for an hour. No Christian should do anything in which he cannot glorify God ; and this may be done, according to Scripture, in eating, and drinking, and the common actions of life. When I have found intense pain relieved, a weary brain soothed, and calm, refreshing sleep obtained by a cigar, I have felt grateful to God and have blessed His name."

" When love grows cool, thy fire still warms me ;
  When friends are fled, thy presence charms me ;
  If thou art full, though purse be bare,
  I smoke, and cast away all care !"

Smoking is a promoter of benevolence. The celebrated German philanthropist, Father Zeller, who was a great smoker himself, said : " When I call upon a man of distinction to ask a favor and I notice a pipe or a cigar-box on the mantel-piece, my hopes rise fifty per cent at once. I am almost sure of success." The use of the pipe he believed to be the emblem of a cheerful, liberal disposition of mind.

IV.

THE ORIGIN OF THE CIGAR—A POPULAR FORM OF THE WEED IN ALL PARTS OF EUROPE AND AMERICA—BISMARCK'S STORY OF CIGAR-SMOKING AT A DIPLOMATIC CONFERENCE—HOW THE SMOKER HAS A GREAT ADVANTAGE IN CONVERSATION—THE DIFFERENT VARIETIES OF CIGARS—CELEBRATED LOVERS OF THE WEED—HOW TO ENJOY A CIGAR—CURIOUS SUPERSTITIONS OF CIGAR-SMOKERS.

The aborigines of America were the first to make tobacco into the rude form of a cigar. Columbus says that the natives rolled the tobacco into a tube or sort of small funnel, formed of the palm leaf, in which the dried leaves of the tobacco were placed; fire was applied to it and the smoke was inhaled. He speaks of this kind of smoking being much used afterwards by captains of ships trading to the West Indies, and says that they attributed to it the power of allaying hunger and thirst, exhilarating the spirits, and renovating the animal powers.

In the narrative of the second voyage of Columbus in 1494, we are informed that the natives reduced the tobacco to a powder, "which they take through a cane half a cubit long; one end of this they place in the nose and the other upon the powder, and so draw it up, which purges them very much." This seems to be the

first notice of snuff-taking ; its effects upon the Indians seem to have been more
violent and peculiar than upon Europeans
since.

In 1699 a traveller, writing of the Indians, says that when the tobacco leaves
are properly dried and cured, the natives,
"laying two or three leaves upon one another, they roll up all together sideways,
into a long roll, yet leaving a little hollow.
Round this they roll other leaves one after
another in the same manner, but close and
hard, till the roll is as big as one's wrist,
and two or three feet in length."

The cigar began to be in vogue among
the Spaniards at the beginning of the seventeenth century. They smoked the leaf
rolled simply or in a leaf of maize, according to the Indian fashion. Although
highly appreciated in Spain, the cigar did
not become acclimated in France much
before 1830. At that period the princes
used to distribute cigars among persons
with whom they wished to be popular.
Oftentimes the recipients of these royal
favors detested tobacco, but felt obliged
to smoke in order to be well at court. The
republic of 1848 showed as much partial-
ity for the pipe as for the cigar, while the
second empire finished by vulgarizing the
cigar. Certain high personalities of that
time made a reputation for themselves as
great smokers, and particularly as con-
sumers of good cigars. The *panatellas* of

Count Jezersky, the *trabucos* of Prince de la Tour d'Auvergne, the *regalias* of Count Cossé, and the *londres* of Prince Serge Gralitzin were renowned.

Modern Spaniards are fond of cigar-smoking. In a Spanish book there is a funny picture of a ball-room scene in Spain, in which there is a fat Spanish countess performing a fandango while she smokes her cigar, of which she is reported to have consumed several during the evening.

The manufacture and use of cigars in Northern Europe only dates from the close of the last century. In 1796 the fashion began in Hamburg and soon spread. Scented cigars were at one time fashionable, and were perfumed with vanilla. German cigars are inferior to the American brand, and are very mild. In Austria and the Italian States their manufacture is a government monopoly.

In Burmah the smoking of cheroots with wrappers made of the leaves of the Then-net tree is very common. In making them, a little of the dried root, chopped fine, is added, and sometimes a small portion of sugar. A traveller says he has seen children two and three years of age, stark naked, tottering about with a lighted cigar in their mouth.

Smoking is a common social custom in Paraguay. Servants in a home bring in a brass vessel, containing a few coals of fire,

and a plate of cigars. Men, women and
children smoke; in the office, the drawing-room, at the dinner-table, and even at
balls and theatres. It is the same in Central America, where every gentleman car-
ries in his pocket a silver case, with a long
string of cotton, steel and flint, and one of
the offices of gallantry is to strike a light;
by doing it well he may kindle a flame in
a lady's heart; at all events, to do it bun-
glingly would be ill-bred.

Sublime tobacco ! which from East to West,
Cheers the tar's labor or the Turkman's rest;
Which on the Moslem's ottoman divides
His hours, and rivals opium and his brides';
Magnificent in Stamboul, but less grand,
Though not less loved, in Wapping or the Strand ;
Divine in hookas, glorious in a pipe,
When tipp'd with amber, mellow, rich and ripe ;
Like other charmers, wooing the caress
More dazzlingly when daring in full dress ;
Yet thy true lovers more admire by far
Thy naked beauties—give me a cigar !

Bismarck, the Prussian statesman, tells
how all the members of the military com-
mission of the Diet at Frankfort took to
smoking at their sittings. At first the
president, Count Rechberg, was the only
one who smoked, until one day Bismarck
coolly asked him for a light and began
smoking a cigar. The other delegates wrote
to their respective governments for instruc-
tions, and the subject, being a grave matter,
required six months for reflection. Mean-
while the Hanoverian representative
smoked, so as to be even with Bismarck,

seeing which, others produced a cigar so
as to be equal with him. The last two
delegates to join the circle of smokers were
not in the habit of using the weed. As
wise diplomats, however, they could not
allow their colleagues to blow clouds in
their faces without blowing back. The
honor of their respective countries was in-
volved. "One of them," says Bismarck,
"brought out an indefinable cigar—pale,
yellow, thin, tapering and enormously
long. He smoked it bravely, with all his
might, and almost to the stump, thus giving a magnificent example of devotion to
his country."

In 1871, at the time of one of the inter-
views between Prince Bismarck and Jules
Favre, the Chancellor began by asking
the French statesman if he would have a
cigar. Jules Favre bowed, and replied
that he never smoked. "You are wrong,"
rejoined Bismarck. "Whenever gentlemen begin a conversation that may sometimes lead to discussions and occasion violent language, it is much better to smoke
while talking. As you smoke," he continued, lighting a fine Havana, "the cigar
that you hold and handle and do not wish
to let fall, paralyzes somewhat the physi-
cal movements. Morally, without depriv-
ing us in any way of our mental faculties,
it lulls us slightly. The cigar is a diversion; the blue smoke which mounts spiral-
ly and that you follow with your eyes in

spite of yourself, renders you more conciliatory. You are happy, your sight is occupied, your hand is retained, and your sense of smell is satisfied. You are disposed to make mutual concessions. Well, our work as diplomatists is made of reciprocal and unceasing concessions. You, who do not smoke, have one advantage over me. You are more wide awake. But you have one disadvantage : you are more inclined to be hasty," he said with a sly smile.

Another famous statesman found the cigar useful in diplomacy. Mazzini, the Italian exile, was forewarned that his assassination had been planned, and that men had been dispatched to London for the purpose, but he made no attempt to exclude them from his house. One day the conspirators entered his room and found him listlessly smoking. "Take cigars, gentlemen," was his instant invitation. Chatting and hesitation on their part followed. "But you do not proceed to business, gentlemen," said Mazzini ; "I believe your intention is to kill me." The astounded miscreants fell on their knees, and at length departed with the generous pardon accorded them, whilst a longer puff of smoke than usual was the only malediction sent after them.

Thackeray says that the man who smokes has a great advantage in conversation. "You may," he says, "stop talking if you like, but the breaks of silence never seem disagreeable, being filled up by the puffing of the smoke ; hence there is no awkwardness in resuming the conversation, no straining for effect, sentiments are delivered in a grave, easy manner. The cigar harmonizes the society, and soothes at once the speaker and the subject whereon he converses. I have no doubt that it is from the habit of smoking that Turks and American Indians are such monstrous well-bred men."

How precious a cigar may be to a smoker is illustrated by an anecdote told by Bismarck himself, who says that at Königgrätz he had only one cigar in his pocket, which he carefully guarded as a miser does his treasure. He looked forward to the happy hour when he should enjoy it, after the battle. "But," he says, "I had miscalculated my chances. A poor dragoon lay helpless, with both arms crushed, murmuring for something to refresh him. I felt in my pockets, and found that I had only gold, which would be of no use to him. But stay, I had still my treasured cigar. I lighted it for him and placed it between his teeth. You should have seen the poor fellow's grateful smile. I never enjoyed a cigar so much as that one which I did not smoke."

Earl Russell was once questioning Tennyson about his visit to Venice. After the poet had said he had seen the Bridge

of Sighs, the pictures and all the wonderful things in the city, the Earl was very
much surprised to hear him say he didn't
like Venice. "How! Indeed! Why
not, Mr. Tennyson?" "They had no
good cigars there, my lord; and I left
the place in disgust."

.The warmth of thy glow,
　Well-lighted cigar,
Makes happy thoughts flow,
　And drives sorrow afar.

The stronger the wind blows,
　The brighter thou burnest!
The dreariest of life's woes,
　Less gloomy thou turnest.

As I feel on my lip
　Thy unselfish kiss,
Like thy flame-color'd tip,
　All is rosy-hued bliss.

No longer does sorrow
　Lay weight on my heart;
And all fears of the morrow
　In joy-dreams depart.

Sweet cheerer of sadness!
　Life's own happy star!
I greet thee with gladness,
　My friendly cigar!

The claro is the mildest grade of cigar;
Colorado claro is the next, then Colorado
maduro, then Colorado medium, and maduro strongest. There are five degrees
of strength, as marked on cigar boxes, in
the ordinary course of trade. There are
certain terms used to describe the shape of

a cigar : coqueta, the smallest; concha, medium ; perfecto, large; Figaro, a shape between coqueta and concha. Invincible are the largest of all, though perfecto cigars are made that have as much tobacco in them. Some invincible cigars are seven inches long. The panatella is two-thirds as long as a lead-pencil, and of about the same diameter. The perfecto is fairly long, big-bellied and usually dark in color. Different manufacturers grade their cigars differently ; the Colorado claro of some makers is as mild as the claro of other makers.

Speaking of mild cigars, it is said that Mazzini had canary birds flying free about his room, and that he always smoked while he wrote. Lord Montairy, in "Lothair," smoked cigars so mild and delicate in flavor that his wife never found him out. Mazzini surely must have had some Montairy cigars, for his canaries did not find him out, or object to him if they did !

The attempts made throughout the world to cultivate the Cuban plant have not given any satisfactory results ; on the other hand, the demand for Havana cigars has increased enormously. As long as the fertile soil of Vuelta Abajo gave sufficiently abundant crops without manuring, the price of tobacco did not advance; but little by little the soil became exhausted, and the consumption steadily increasing, the planters used strong fertilizers, such as guano and house refuse. The result of this intense cultivation has been satisfactory for the Cuban planters, but unfavorable for the smokers. It is certain that three-quarters of the cigars sold as Havanas do not contain any Vuelta Abajo tobacco, or at least only the remains of the bad leaves. Cigar-making has reached such a perfection in the United States that it is very difficult to distinguish the imported cigar from the domestic, except by trying it; and in many cases the most experienced smoker can scarcely tell the difference.

In Spain, the Seville manufactory has acquired a European reputation for the making of high-priced cigars, and produces an article equal in appearance to the finest Havana. In Belgium and at Hamburg and Frankfort, cigars are made of beet-root leaves steeped in a decoction of tobacco juice, and sold as pure Havanas. Germany, for that matter, excels in counterfeiting Havana cigars. The manufacturers there take a poor quality of Virginia or Rhine tobacco as a filling, and cover it with a magnificent wrapper. As soon as a vessel from Cuba is signalled off Hamburg or Bremen, thousands of these bogus Havanas, all packed in boxes, marked and ribboned, as though made at the Gem of the Antilles, are put on board. When the ship reaches her dock these cigars are entered at the Custom House as coming from Havana. The cedar-wood, the paper, and even the little nails used in the manufacture of the boxes, are sent to Germany by the Cuban merchants.

To-day, as half a century ago, cigars are made by hand, for no one has yet been able to invent a machine that will roll a cigar with the same care as a woman's fingers. In France alone, more than 17,000 women are employed in the Government tobacco manufactories, and a good hand can roll from 100 to 150 choice cigars in ten hours. These women are not allowed to speak during working time, but when they leave the factory they make up for lost time. In the cheaper cigars, French and foreign tobaccos are always more or less mixed, the proportions being variously regulated. The ordinary one and two-cent cigars are made of French, Kentucky, Algerian or Hungarian leaves. All the cigars sold at ten cents and above are bought directly by the French Régie from the Havana manufacturers. These cigars are, upon their arrival, sent to the Government factory in the quarter known as Gros Caillou, where they are unpacked and examined to see if they have arrived in good condition. To make sure, three inspectors take a handful here and there and smoke them, not for themselves, but for the public. If the experts find that the cigars have lost any of their qualities of taste or flavor they reduce the price,

and if the change is too marked, the lot is
shipped to some foreign country and sold at
the best price attainable. Sometimes these
damaged cigars are smuggled back to
France and sold at high prices. And the
flats who buy and smoke them with delight
exclaim : "If the Régie would only furnish us such cigars !"

Very expensive cigars are bought by the
aristocracy of Europe—princes and kings
principally. The Czar of Russia smokes
a dozen $1.50 cigars a day. In former
times the best Havana tobacco leaves were
reserved for cigars for the King of Spain,
and one particularly large and fine kind of
cigar was used especially by the priests ;
such being made from the picked leaves
which were presented to the Church and
manufactured by the monks themselves.

In a play written in the seventeenth century the hero says : "Look at me—follow
me—smell me ! The 'stunning cigar' I am
smoking is one of a sample intended for
the Captain-General of Cuba, and the King
of Spain, and positively cost a shilling !
Oh ! I have some dearer at home. Yes,
the expense is frightful, but who can
smoke the monstrous rubbish of the
shops ?"

CONFESSION OF A CIGAR-SMOKER.

I owe to smoking, more or less,
Through life the whole of my success ;
With my cigar, I'm sage and wise—
Without, I'm dull as cloudy skies.

When smoking all my ideas soar,
When not, they sink upon the floor.
The greatest men have all been smokers,
And so were all the greatest jokers.
Then ye, who'd bid adieu to care,
Come here and smoke it into air.

Richard Porson, the celebrated Greek
scholar, was not only very fond of alcohol-
ic stimulants, but consumed prodigious
quantities of tobacco. On one of his or-
gies, which he would indulge in after
weeks of unremitting labor, he emptied a
half-pound canister of snuff, and in one
night smoked a large bundle of cigars.
"Previous to this exhibition," said the host
who had entertained him, "I had always
considered the powers of man limited."

Mr. Goodman, an Englishman well
known in turf circles, in 1860, on a wager,
smoked one pound of strong foreign regalias within twelve hours. The cigars ran
eighty-six to the pound, so that the smoker
consumed eight an hour. He commenced
his task at 10 A. M. and finished at
7.20 o'clock P. M. In the course of nine
hours and twenty minutes seventy-two
cigars were fairly smoked out, the greatest
number consumed being in the second
hour, when the smoker disposed of no less
than sixteen. At the seventy-second cigar,
when fourteen only remained to be
smoked, the backer of time gave in, find-
ing that Mr. Goodman was sure to win.
The smoker declared that he felt no un-
pleasantness during the task. The only

refreshment taken was a chop at two
o'clock, and two-thirds of a pint of brandy
in cold water at intervals during the
smoking.

A regular smoker in Cuba will consume
perhaps twenty or thirty cigars a day, but
they are all fresh. What we call a fine
old cigar, a Cuban would not smoke.

Girardin was a great smoker. Charles
Dickens met him in Paris, and says that
after dinner the Frenchman asked him if
he would not step into another room and
smoke a cigar. After entering the apartment, Girardin coolly opened a drawer,
containing about 5000 inestimable cigars
in prodigious bundles ; just as the captain
of the robbers in Ali Baba might have
gone to the corner of the cave for bales of
brocade.

Nearly all literary men have been friendly to tobacco. Jules Sandeau says that
the cigar is one of the greatest triumphs of
the old world over the new. It is an in-
dispensable complement of all idle and
elegant life, and the man who does not
smoke cannot be regarded as perfect. He
says that the cigar of to-day has taken the
place of the little romances, coffee and
verses of the seventeenth century. Spain,
Turkey and Havana have yielded up to us
the most precious treasures of their smoke-
enwrapt dreamland. Speaking of the
charming reveries that come to the cigar-
smoker, he says : " Let me tell you, that if

you have never found yourself extended upon a divan with soft and downy cushions on some winter's evening before a clear and sparkling fire, enveloping the globe of your lamp or the white light of your wax candle with the smoke of a well-seasoned cigar, letting your thoughts ascend as uncertain and vaporous as the smoke floating around you, let me tell you, I repeat, that if you have never yet enjoyed this situation, you have still to be initiated into one of the sweetest of our terrestrial joys. The cigar deadens sorrow, distracts our enforced inactivity, renders idleness sweet and easy to us, and peoples our solitude with a thousand gracious images. Solitude without friend or cigar is indeed insupportable to those who suffer. It is through the fragrant weed that we drift into indolence, and become dreamy, contemplative, useless creatures. Thackeray called the cigar the greatest creature-comfort of his life—a kind campanion, a gentle stimulant, an amiable anodyne, a cementer of friendship.

To enjoy a cigar, according to epicurean fashion, the end should be cut smoothly off by the clipper, the cigar should be blown through for the purpose of removing all the little particles of dust which cannot be avoided in manufacture ; this prevents them from being inhaled into the throat and from producing coughing. The cigar should then be lighted—thoroughly lighted all over the surface of the end. Three or four puffs every minute will enable one to enjoy the smoke. The smoke should be kept in the mouth a short time in order to appreciate the flavor. Then it should be emitted slowly. In case one side of the cigar should burn and leave a ragged edge on the other side, a gentle blow through the cigar toward the lighted end will ignite the ragged side and it will burn regularly. If a cigar is smoked in this way, it is a pleasure.

A man's disposition is shown by the way he smokes a cigar. Tranquil men smoke a cigar without the ashes falling off. A nervous man taps with his little finger on the cigar, or the motions of his hand will cause the ashes to fall off. Some men smoke a cigar steadily and evenly, others make it ragged and light it several times in the course of a conversation.

If a man smokes his cigar only enough to keep it lighted, and relishes taking it out of his mouth to watch the curl of the smoke in the air, he may be set down as an easy-going man. The man who never releases his grip on the cigar is cool, calculating and exacting. The man who smokes and stops alternately is easily affected by circumstances. The man whose cigar goes out frequently is of a whole-souled disposition. The man who "monkeys" with his cigar is a sort of popinjay among men. The fop stands his ci-

66

gar on end, but the experienced smoker
points it straight ahead or almost at right
angles with his course. The question has
often been raised among smokers as to
when a cigar tastes best. This can only
be decided by each smoker for himself;
but nearly all lovers of the weed enjoy a
smoke after eating a meal.

An epicure, prominent in one of the New
York clubs, says that a cigar tastes best in
the morning. The reason for this is, that
the man at that time is fresh and invigo-
rated. If a man smokes many cigars in a
day, he cannot enjoy them all equally well.
It is like taking too many cocktails before
dinner; the man who does that cannot
appreciate the best effects of a good cook
or the delicate bouquet and flavor of a
fine Burgundy.

Cigar-smokers have certain superstitions
for which they cannot very well give
reasons. Some do not believe that a man
should smoke after breakfast or immediately before meals; others think that a
cigar that has once gone out does not
smoke so well as a cigar burned through
steadily; that the last inch of a cigar is
the best, and that the strength of a cigar
is determined by the color of the wrapper.
Smokers do not stop to consider that the
wrapper forms but a small part of the
bulk of a cigar, and that its strength or
mildness is determined by the filler and
not by the wrapper. There is a prejudice

in the minds of most smokers against smoking a cigar that has once gone out; but the fact that half an inch of a cigar has been smoked does not necessarily make the rest of it worthless.

A young man once consulted the famous Dr. Abernethy. After interrogating the patient upon his life and habits, Abernethy was puzzled to account for the state in which he found the sufferer; suddenly a thought struck him. "Do you expectorate, sir?" he enquired. The patient replied that since he smoked a good deal, spitting had become habitual to him. "Ah! that need not cause you to expectorate," mused the doctor. "Well, well," he resumed, "I'll just take time to think over your case; you can call on me to-morrow morning, at eleven o'clock, for a prescription." The following morning, Dr. Abernethy's patient punctually made his appearance. "I'm very sorry, sir, but I have a pressing engagement just now; if you'll step upstairs into my drawing-room and wait for half an hour, you'll find a box of Colorados to amuse yourself with." "Well, now, what do you think of my cigars?" said Abernethy, when, in the course of an hour, he came into the room in which his patient awaited him—a room, be it said, luxuriously furnished with every possible convenience except that of a spittoon. "I enjoyed the first so much that I could not help taking a second." "But where, then," said the doctor, prying curiously under the table and inside the grate, "have you been spitting ?" " Good gracious, doctor, what can you be thinking of, to imagine that, in such a place, I should do otherwise than swallow my spittle !" "Pay me my fee," said the doctor, "and go, and remember I never say that you cannot smoke without spitting. That is your sole complaint."

A cigar is good even after it ceases to be a cigar. It is is said that cigar ashes mingled with camphorated chalk make an excellent tooth-powder ; or ground with poppy-oil, will afford for the use of the painter a varied series of delicate grays. Old Isaac Ostade so utilized the ashes of his pipe ; but had he been aware of Havanas, he would have given us pictures even more pearly in tone than those which he has left for the astonishment and delight of mankind,

V.

How Pipes are Made—Curious Snuff-
Boxes—Snuff - Taking in England,
Spain, Italy, and France — Famous
Lovers of Snuff—Rare Collections
of Remarkable Snuff-Boxes — How
to Take a Pinch of Snuff.

A curious old pipe-maker in New York,
an Austrian by birth, boasts of having
served a long apprenticeship at his trade,
and says that he passed six examinations in
his profession, in amber, meerschaum, rub-
ber, ivory, wood and metals. He is an
adept in carving, and has made an amber
skull less than three quarters of an inch in
height in which the bones and articulations
are distinctly marked. The carving is so
fine that a magnifying glass has to be used
to see it in detail. Another is a holder,
where a monk with a hollow head for
cigarettes is laughing, but it requires a
magnifying glass to see the lines of his
mirth. Another design is a wine bowl in
form of a skull, hollowed out for
cigarettes. The most costly pipe represents a mermaid holding a sea shell close
to her breast ; her scaly tail is twined about
a large branch of white coral, which becomes brown when the pipe is smoked.

When an order comes for a pipe the
proprietor selects from his stock of meerschaum a piece from which it can be cut
with as little loss as possible. Four-fifths

of the meerschaum is wasted, though the chips are often saved and made into imitation meerschaum pipes. The meerschaum is first cut on a circular saw into a piece a little larger than the pipe. If the cutting shows cracks or holes it is cast aside. Then it is soaked in water for fifteen minutes and cut the rough shape with a knife. Then a hole is drilled through it and it is turned, after which the stem is inserted. It is smoothed off when dry, boiled in wax, and polished, and is then ready for the market.

The amber is worked with a razor-like chisel and turning wheel. After being rounded it is held against the face of a roughened wheel until it is made to approximately the required size. Then a hole is bored through it. This is the process for the cheaper amber stems, which can be made in a quarter or half an hour; a stem for a costly pipe will take a day. It takes three days to make a good, plain meerschaum pipe, but a carved pipe may require several months. The dust and chips from the amber and meerschaum are saved; the amber dust is melted and made into amberine, and the meerschaum dust is made into a paste from which imitation meerschaum pipes are made.

Quaint forms are as common to snuff-boxes as to tobacco-pipes. One favorite in the last century was a lady's shoe, carved in wood and inlaid with threads of silver

to imitate ornamental stitches. Coffins
were also hideously adapted to hold the
fragrant "dust." A coiled snake, whose
central folds form the lid, was a box for a
naturalist ; a book might serve for a student, and a boat for a sailor. All persons
and all states may be "fitted" with a
proper receptacle for the pungent dust they
love so well, and of which the rhymester
sings :

" What strange and wondrous virtue must there be,
And secret charm, O snuff, concealed in thee !
That bounteous Nature and inventive Art,
Bedecking thee, thus all their powers exert ;
Their treasures and united skill bestow
To set thine honours in majestic show !
But oh ! what witchcraft of a stronger kind,
Or cause too deep for human search to find,
Makes earth-born weeds imperial man enslave,
Not little souls, but e'en the wise and brave !"

Gillespie, an Englishman, who made a
fortune out of making a snuff which bore
his name, had this motto for the arms
on his carriage :

" Who could have thought it
That noses had bought it ?"

The Scotch were such large snuff-takers
that the figure of a Highlander helping
himself to a pinch was used as a sign by
the snuff-shops. An old Highlander in
urging a friend to visit him says :

" There'll be plenty of pipe, and a glorious supply
Of the good sneesh-te-bacht, and the fine cut and
dry ;

There we'll drink foggy Care to his gloomy abodes,
And we'll smoke, till we sit in the clouds, like the
gods.

Snuff was first used medicinally, particularly for diseases of the head brought on
by colds. Catherine de Medicis was the
first so to use it in the court of France,
about 1562. An old English doctor in
1610 recommended snuff ; "being drawne
up into the nostrels, cause sneesing, consuming and spending away grosse and
slimie humors from the ventricles of the
braine." Another method of using it was
to make it into small suppositories, or pel-
lets, and put them up into the nose. The
Irish were remarkable snuff-takers. A
writer in 1659 informs us : "The Irish are
altogether for snuff tobacco to purge their
brains."

During the early part of the seventeenth
century taking a pinch of snuff was common in Spain, Italy and France. Pope
Innocent XII., in 1690, excommunicated
those who should take snuff or tobacco in
St. Peter's at Rome. But the prelates and
religious community were fond of it, in
spite of the Pope and his ordinances, and a
writer of those days says, "The Spanish
priests will not scruple to place their snuff-
boxes on the altar for their use."

At this time tobacco was reduced to a
rough powder by pounding or grating.
In America the tobacco was laid away in
twisted rolls and taken out, as occasion re-

quired, for the purpose of being made into smoking tobacco, chewing tobacco, or snuff. When snuff was made a quantity was taken from the roll and laid in a room where a fire was kept. In a day or two it would be dry, and was rubbed on a grater, producing a genuine snuff. Sometimes it was scented by the use of odoriferous waters.

The outfit of a fashionable snuff-taker at this period was quite costly, and the tobacco-grater, formed of ivory, was richly carved with a variety of scroll ornament enclosing fanciful scenes of various kinds. The snuff-grating machine was very much like the ordinary grater used to grate nutmegs. Some manufacturers pounded the leaves in a mortar, the pestle being of peculiar form to allow the more perfect mixing of the scents so commonly used.

Scented snuff gave a chance for a gallant to pay a compliment :

> " Dear Jenny, if this snuff should want
> Such odours as your breath bestows,
> Your touch will give 't a sweeter scent
> Than quintessence of fragrant rose."

When Dryden frequented Will's coffee-house it became a great resort of the wits of his time. A newspaper writer of that period says that "a parcel of raw, second-rate beaux and wits were conceited if they had but the honor to dip a finger and thumb into Mr. Dryden's snuff-box."

Frederick the Great loved it so well that he carried it in capacious pockets made in his waistcoat, that he might have as little trouble as possible in getting at it. Dr. Johnson was probably a snuff-taker of this kind. George II. and Napoleon carried snuff in a similar way. Many of the sovereign pontiffs of the Roman Catholic Church have been confirmed snuff-takers.

So common was the practice in France in 1774 that persons distributed boxes of snuff to passengers as they crossed the bridge in Paris. This was a scheme to introduce it into general use. At this period, an old French writer asserts, there was no person in France, of whatever age, rank, or sex, that did not take snuff.

In an English satire, written in 1710, the hero's snuff-box is described as being filled with a snuff called Orangery: "After dinner the ladies, all impatient for the first pinch, put in their fingers almost all at once ; the gentlemen with some respect after."

Addison, in his *Spectator*, put this pertinent inquiry to the beaux of this period : "Would it not employ a beau prettily, if, instead of playing eternally with a snuff-box, he spent some part of his time in making one ?"

"Knows he that never took a pinch,
  Nosey, the pleasure thence which flows ?
 Knows he the titillating joys.
      Which my nose knows ?

O nose ! I am as proud of thee
As any mountain of its snows ;
I gaze on thee, and feel that pride
A Roman knows."

Snuff graters went out of use long ago,
and are now to be found only in museums
or private collections. However, as late
as 1820 there were in France official snuff
graters, men who travelled from chateau
to chateau, and from parsonage to parsonage to pulverize the tobacco of the priest
or the dowager. The graters were made
of wood, ivory, brass, iron, etc., and often
carved on one side in the most elaborate
manner ; on the other side were the little
holes through which the powder fell as
the roll of tobacco was rubbed over them.
Besides the simple graters there were
others, called *grivoises*, surmounted by
snuff-boxes. Until the end of the
eighteenth century these snuff graters were
used exclusively by the upper and richer
classes. The common people, who could
not afford the luxury of a grater, or even
buy their tobacco by the pound, were
obliged to content themselves with the
snuff sold at the street corners, which was
often adulterated with powdered glass " to
make it more stimulating."

According to all probability, snuff was
introduced into the Orient in the seventeenth century. In China the snuff-takers are less numerous than the smokers.
" Smoke for the nose," as the Celestials

call snuff, is but little used except among
the Mantchoo Tartars and Mongolians, and
only by the lettered class and the manda-
rins. The best snuff, called *piyinn*, is
made at Canton, and is rare. Ordinary
snuff is sent from Portugal and Spain to
Macao. The Chinese preserve their snuff
in little bottles of crystal, porcelain, or of
precious stones. and wear them fixed to
their belts. Attached to the stopple by a
little chain is a small spatula in ivory or
silver, which they use for taking the snuff
out of the bottle. Then they place the
snuff on the back of their left hand. near
the last thumb joint, and inhale it slowly,
with a sort of amorous pleasure. The
Japanese take their snuff in the same way.
A like usage exists in India. In that
country the snuff-boxes are made with
gourds, cocoanuts, and buffalo horns. In
Van Diemen's Land the inhabitants use
iron wood, Huron pines, musk wood,
whales' teeth; etc. Among the Turks
there are many snuff-takers, and the habit
is also prevalent with the Afghanistans.

Tobacco reached the height of its honors
at the time of the appearance of snuff-
boxes. From the court of France and the
nobility they passed into the hands of
everybody during the second half of the
eighteenth century. They were of all
shapes and kinds, from the commonest
wood to the most costly materials. The
eighteenth century was, indeed, the cen-

tury of the snuff-box ; not a single nose of grand seignior, peasant, marquis, or ballet girl escaped its domination. At first the portraits on the snuff-boxes were placed on the inside of the cover, and at the end of a few days the painting turned yellow and became almost effaced. An idea of mystery was evidently the cause of this custom. The grand seigniors used to wear their snuff-boxes as jewels, and in their houses they displayed them in glass cases and on the mantels. Naturally, these boxes soon got to be the fashionable present, and were offered for all sorts of reasons. Marie Antoinette received fifty-two golden ones at her marriage.

During the revolutionary period, and down to 1830, the snuff-box became a political instrument, and was used as a sign of recognition among the conspirators of the different parties. There were snuff-boxes called the "Bastille," the "Mirabeau," the "Bonnet Phrygien," the "Martyr of Liberty," the "Rat Tail," "Madame Angot," etc. One of the most precious ones existing was given to Danton at the time of his marriage by Camille Desmoulins ; it now belongs to M. Spuller, the ex-Minister. From the Consulate, down to the time of the death of Napoleon I., the snuff-boxes reproduced his features, those of his family, his Generals, and the illustrious men of his time. The Emperor made presents of valuable snuff-boxes en-

riched with diamonds, while those that
he carried were simple, narrow, oval boxes
in black shell, lined with gold and ornamented with cameos or antique medallions
in silver. One of the rarest snuff-boxes of
that period is that given by Pope Pius
VII. to Napoleon at the time of his coronation. The little cocked hat that the Emperor wore also gave its shape to one of
the most popular of snuff-boxes, but it was
proscribed during the Restoration. After
1830 the round box with the cover ornamented with portraits or emblems disappeared and was replaced by the large flat
boxes and hinged cover. This shape, made
of all sorts of material and more or less
ornamented, is still used.

As for the collections of snuff-boxes,
they are numerous. The Prince de Conti,
who died in 1776, left 800. Frederick the
Great is said to have had even more than
this number ; snuff-boxes were his greatest hobby. The Duke of Richelieu had
one for each day in the year. The
Regent's collection was also celebrated ;
it remained in the Orleans family until
1848, when it was sold at auction. The
Princesse de Tallard, governess of Louis
XV.'s children—the legitimate ones—possessed a remarkable collection. The
Fermier-General Pinon, Vigée, the poet ;
Lablache, the singer, and the Prince
Demidoff were celebrated collectors. Of
a more recent date, two collections are

worthy of mention : the one left by Mme. Lenoir to the Louvre, in 1874, consisting of 204 boxes in gold, ornamented with paintings, enamels, and precious stones ; the other belonging to M. Alphonse Maze-Sencier, contains a series of all shapes, from the eighteenth century to the end of the Second Empire, and forms the most complete history that exists upon the subject.

Talleyrand was a snuff-taker, not from devotion to the habit, but on principle. The wily politician used to say (and doubt-less Metternich, who was a confirmed snuff-taker, would have agreed with him) that all diplomatists ought to take snuff, as it afforded a pretext for delaying a reply with which one might not be ready ; it sanctioned the removal of one's eyes from those of the questioner ; occupied one's hands which might else convict one of nervous fidget ; and the action partly concealed that feature which is least easily schooled into hiding or belying human feelings—the mouth. If its workings were visible through the fingers, those twitches might be attributed to the agreeable irritation going on above.

No other article of *vertu* has been more extensively patronized by the crowned heads of Europe, for purposes of presentation, diplomatic or otherwise, than the snuff-box. In evidence of its importance as a means of keeping up friendly relations with foreign powers, we need only

quote, from the account of sums expended
at the coronation of George IV., the following entry : Messrs. Randell & Bridge,
for snuff-boxes to foreign ministers, £8205
15s 5d.

Gibbon, the historian of Rome, was a
confirmed snuff-taker, and in one of his let-
ters has left this account of his mode of
using it : " I drew my snuff-box, rapp'd
it, took snuff twice, and continued my
discourse in my usual attitude of my body
bent forward, and my forefinger stretched
out." In the *silhouette* portrait he is represented as indulging in this habit, and
looking, as Colman expresses it, " like an
erect, black tadpole, taking snuff."

The successful strategist Count Moltke
is an inveterate snuff-taker. In the grand
three weeks' campaign which culminated in
that Prussian " Waterloo," the battle of
Sedan, his plans were assisted by a pound
of snuff. Throughout the Prussian advance, amid its tremendous anxieties, the
General took · snuff to excess, but at the
supreme moment when the Uhlans an-
nounced to him the march northward of
Marshal MacMahon, Moltke literally emptied his snuff-box as he entered his tent
to organize the movement which resulted
in the capture of Napoleon III. on the Bel-
gian frontier. And strange to tell, adds
Mr. Steinmetz, Moltke was actually required, by the German War-Office, to pay
for that memorable pound of snuff at the

end of the war, when there was presented
to him the bill (duly signed and counter-
signed by various officials), which ran,
" For one pound of snuff supplied to General Von Moltke, one thaler !"

As already stated, Frederick the Great
took large quantities of snuff. To save
himself the trouble of extracting it from
his pocket, he had large snuff-boxes placed
on each mantel-piece in his apartments, and
from these would help himself as the fancy
took him. One day he saw, from his
study, one of his pages, believing himself
unobserved, put his fingers unceremoni-
ously into the open box on the adjoining
mantel-piece. The King said nothing at the
moment, but after the lapse of an hour he
called the page, made him bring the snuff-
box, and bidding the indiscreet youth take
a pinch from it, said to him, " What do
you think of the snuff ?" " Excellent,
sire." "And the box ?" "Superb,
sire." " Oh, well, sir, take it, for I think
it is too small for both of us !"

The cynical temper of Frederick the
Great is well known. He once made a present of a gold snuff-box to the brave Count
Schwerin. Inside the lid the head of an
ass had been painted. Next day, when
dining with the King, Schwerin ostenta-
tiously displayed his snuff-box. The
King's sister, the Duchess of Brunswick,
who happened to be staying at Potsdam,
took it up and opened it. Immediately she

exclaimed, " What a striking likeness ! In
truth, brother, this is one of the best portraits I have ever seen of you." Frederick,
much embarrassed, thought that the Duchess was carrying the joke too far. She,
however, passed the box to her neighbor,
who gave vent to similar expressions of as-
tonished admiration. The box made the
round of the table, and every tongue
waxed eloquent upon the subject of this
" counterfeit presentment." The King was
extremely puzzled, but when the box at
length reached his own hands, he saw, to
his great surprise and greater relief, that
his portrait was indeed really there. The
wily Count had simply employed an artist
to remove with exceeding despatch the
ass's head, and substitute for it the King's
well-known features. His Majesty could
not but laugh at the clever device which
had so completely turned the tables on him.

Robert Burns was never happier than
when he could " pass a winter evening
under some venerable roof and smoke a
pipe of tobacco or drink water gruel."
He also took it in snuff. Mr. Bacon, who
kept a celebrated posting-house north of
Dumfries, was his almost inseparable asso-
ciate. Many a merry night did they spend
together over their cups of foaming ale or
bowls of whiskey-toddy, and on some of
those occasions Burns composed several of
his best convivial songs. The bard and the
innkeeper became so attached to each other

that, as a token of regard, Burns gave
Bacon his snuff-box, which for many years
had been his pocket companion. The
knowledge of this gift was confined to a
few of their jovial brethren. But after
Bacon's death, in 1825, when his household
furniture was sold by public auction, this
snuff-box was offered among other trifles,
and some one in the crowd at once bid a
shilling for it. There was a general ex-
clamation that it was not worth twopence,
and the auctioneer seemed about to knock
it down. He first looked, however, at the
lid, and then read in a tremendous voice the
following inscription upon it: "Robert
Burns, officer of the Excise." Scarcely
had he uttered the words, says one who
was present at the sale, before shilling after
shilling was rapidly and confusedly offered
for this relic of Scotland's great bard, the
greatest anxiety prevailing; while the bid-
dings rose higher and higher, till the trifle
was finally knocked down for five pounds.
The box was made of the tip of a horn,
neatly turned round at the point; its lid is
plainly mounted with silver, on which the
inscription is engraved.

Speaking of Scotland, there is a story of
the snuff-mull in the Scotch kirk. An
English lady found herself in a parish
church not far from Craithie, in a large
pew occupied by farmers and their wives
and one or two herdsmen—about a dozen
in all. Just before the commencement of

the sermon a large snuff-mull was handed
round ; and upon the stranger declining to
take a pinch, an old shepherd whispered
significantly, "Tak' the sneeshin', mem ;
tak' the sneeshin'. Ye dinna ken oor minister ; ye'll need it afore he's dune."

Here is a pen-picture of the famous
Bishop Whately as a snuff-taker :—The
logic class is assembled. The door by
which the principal is to enter is exactly
opposite to the foot of the stair which descends from his own apartment. It stands
open, and presently a kind of rushing sound
is heard on the staircase. The next instant, Whately plunges headforemost into
the room, saying while yet in the door-
way, " Explain the nature of the third operation of the mind, Mr. Johnson." But
as none of the operations of Mr. Johnson's
mind are so rapid as those of the energetic
principal, the latter has had time to fling
himself into a chair, cross the small of one
leg over the knee of the other, balance himself on the two hind legs of the chair, and
begin to show signs of impatience, before
Mr. Johnson has sufficiently gathered his
wits together. While that process is being
accomplished, the principal soothes his impatience by the administration of a huge
pinch—or handful, rather—of snuff to his
nose, copiously sprinkling his waistcoat
with the superfluity thereof. Then at last
comes from Mr. Johnson a meagre answer
in the words of the text-book, which is fol-

lowed by a luminous exposition of the rationale of the whole of that part of the subject, in giving which the lecturer shoots far over the heads of the majority of his hearers, but is highly appreciated by the select few who are able to follow him.

Directions for taking a pinch of snuff :— The true snuff-taker, who is bold in his propensities, always has a large wooden snuffbox, which he opens with a crash, and which he flourishes about him, with an air of satisfaction and pride. He takes a pinch with three fingers, and then, bringing the whole upon his thumb, he sniffs it up with that lusty pleasure with which a rustic smacks a kiss upon the round and ruddy cheek of his sweetheart.

The true artistic method, however, of "taking a pinch" consists of twelve operations :—

1. Take the snuff-box with your right hand.

2. Pass the snuff-box to your left hand.

3. Rap the snuff-box.

4. Open the snuff-box.

5. Present the box to the company.

6. Receive it after going the round.

7. Gather up the snuff in the box by striking the side with the middle and forefinger.

8. Take up a pinch with the right hand.

9. Keep the snuff a moment or two between the fingers before carrying it to the nose.

10. Put the snuff to your nose.

11. Sniff it in with precision by both nostrils, and without any grimace.

12. Shut the snuff-box, sneeze, spit, and wipe your nose.

A spectator in the pit at the Opera felt a certain pressure upon his coat-pocket, cf the aim and object of which he was but too well aware. "You have taken my snuff-box," said he quickly but cautiously to an individual, of very suspicious aspect, who was standing next him. "Return it to me, or I—" "Don't make a noise, I beseech you ; pray don't ruin me. Here, take back your snuff-box," added the shabby customer in a low voice, at the same time holding his coat-pocket wide open, into which the too confiding owner of the missing *tabatière* thrust his hand. The rogue immediately caught hold of it and cried "Thief ! thief !" and showed the imprisoned hand to the spectators. The veritable owner of the snuff-box was forthwith arrested, but, of course, soon proved his innocence. In the mean time, however, both snuff-box and accuser had disappeared !

VI.

THE HABIT OF CHEWING TOBACCO, OR A
SUBSTITUTE COMMON AMONG ALL UN-
CIVILIZED RACES—" CHEWING" IN PAR-
AGUAY, IN THE FAR EAST, LAPLAND, THE
CAPE OF GOOD HOPE, ETC.—A CURIOUS
CALCULATION FOR TOBACCO-CHEWERS
AND TEA-DRINKERS—A DISTINGUISHED
CLERGYMAN'S DEFENCE OF CHEWING—
" DON'T FORGET THE PIG-TAIL"—PIPE-
SMOKING IN FRANCE, HOLLAND, GER-
MANY, SPAIN AND ITALY.

The Malays are fond of a narcotic, and
the indulgence in opium is not unknown,
but the national indulgence of the race is
the areca, or betel-nut, a habit characteris-
tic of a sea-loving people. The use of a
pipe, especially an opium pipe, would be
a hindrance to the freedom of their motions on board their vessels, and require a
state of inactivity or repose incompatible
with a maritime life, in order to be enjoyed. This may in part account for the
prevalence of chewing tobacco in our navy
and the nut-chewing habit of the Malays.

In Paraguay everybody smokes, and
nearly every woman and girl more than
thirteen years old chews tobacco. A magnificent Hebe, arrayed in satin and flashing
in diamonds, puts you back with one delicate hand, while with the fair taper fingers
of the other she takes the tobacco out of
her mouth previous to your saluting her.

In Siberia boys and girls of nine or ten years of age put a large leaf of tobacco into their mouths without permitting any saliva to escape, nor do they put aside the tobacco should meat be offered to them, but continue consuming both of them together.

The Mintria women and other races of the great Indian Archipelago are addicted to chewing tobacco. Among the Nubians the custom is more common than smoking.

The Finlander delights in chewing. He will remove his quid from time to time, and stick it behind his ear, and then chew it again. This reminds us of a circumstance narrated by a friend, which occurred when he was a boy. His master was a chewer. After a " quid " had been masticated for some time it was removed from his mouth and thrown against the wall, where it remained sticking ; the apprentice was then called to write beside it the date at which it was flung there, so that it might be taken down in its proper turn, after being thoroughly dried, to be chewed over again.

At the Cape of Good Hope grows a plant allied to the ice-plant of our greenhouses, and which is a native of the Karroo, which appears to possess narcotic properties. The Hottentots know it under the name of Kow, or Kauw-goed. They gather and beat together the whole plant, roots, stem, and leaves, then twist it up like pig-tail tobacco, after which they let the mass ferment and keep it by them for chewing, especially when they are thirsty. If it be chewed immediately after fermentation it is narcotic and intoxicating. It is called canna-root by the colonists.

In Lapland, Angelica-root is dried and masticated in the same way, and answers the same purpose as tobacco. It is warm and stimulating, and not narcotic, nor does it leave those unpleasant and unsightly evidences of its use which may be observed about the mouth of the true votary of the quid.

The Duke of Marlborough has the credit of being the first distinguished man who made the chewing of tobacco famous.

Somebody with a strong antipathy to pig-tail and fine-cut has entered into certain investigations and calculations in this wise : If a tobacco chewer chews for fifty years, and uses each day of that period two inches of solid plug, he will consume nearly one mile and a quarter in length of solid tobacco half an inch thick and two inches broad, costing $2094. By the same process of reasoning, this statist calculates that if a man ejects one pint of saliva per day for fifty years, the total would swell into nearly 2300 gallons—quite a respectable lake.

Another calculation shows that if all the tobacco which the British people consumed during three years were worked up into pig-tail half an inch thick, it would

form a line 99,470 miles long, or enough
to go nearly four times round the world ;
or if the tobacco consumed by the same
people in the same period were to be placed
in one scale and St. Paul's Cathedral and
Westminster Abbey in the other, the eccle-
siastical buildings would kick the beam.
Let us compare therewith the tea-consump-
tion during the past three years. There
were consumed about 205,500,000 pounds
of tea, which, if done up in packages containing one quarter of a pound each (such
packages being $4\frac{1}{2}$ inches in length and $2\frac{1}{2}$
inches in diameter), these placed end to end,
would reach 59,428 miles ; or, upon the
same principles as those adopted for the
pig-tail, would girdle the earth twice with
a belt of tea $2\frac{1}{2}$ inches in diameter, or
twenty-five times that of the aforesaid pig-
tail—enough to make rivers of tea strong
enough for any old lady in the kingdom to
enjoy, and deep enough for all the old
ladies in the kingdom to bathe in.

When Rev. Dr. Tiffany, of Minneapolis.
preached in Chicago, his brethren all knew
that he loved fine-cut, because he made no
secret of the chewing habit. He was a
regular attendant at those Monday morning "ministers' meetings" which the
average reporter hates, but which are
really enjoyable on account of the bright
sayings and clever witticisms of preachers
who do not think they are forbidden to indulge in a hearty laugh because they oc-

cupy a pulpit. While Dr. Tiffany was a participant in these meetings the tobacco habit came up for discussion one morning. A well-known bishop was presiding. One after another the brethren arose and con-- demned the use of tobacco in any form. Then one of them, during a lull, said he would like to hear Dr. Tiffany's ideas on the subject. The big doctor arose. " I chew tobacco," he said, "and you all know it. Now I would like to have all those who do not use tobacco rise in their seats." There was a grand uprising. " Remain standing, please," said the doc- tor, as he looked over the cadaverous men standing before him. " Will those who use tobacco please step forward here ?" he said, and a half dozen sleek-looking par- sons walked up and joined him. " Stand up, bishop ; you're a chewer," he said to the presiding divine, and he joined the group. Dr. Tiffany then looked over the thin fellows who tabooed tobacco, turned to the healthy looking men around him, and said : " Brethren, I think we are doing pretty well." The argument was un- answerable.

The following letter, written by a sailor, aptly illustrates the attitude of Jack Tar toward chewing tobacco :

GRAVESEND, Mar. 24, 1813.

DEAR BROTHER TOM : This comes hop- en to find you in good health as it leaves

me safe ankor'd here yesterday at 4 P.M.
arter a pleasant voyage tolerable short
and a few squalls. Dear Tom,—hopes to
find poor old father stout, and am quite
out of pig-tail. Sights of pig-tail at Gravesend, but unfortinly not fit for a dog to
chor. Dear Tom, Captain's boy will bring
you this, and put pig-tail in his pocket
when bort. Best in London at the Black
Boy in 7 diles, where go acks for best
pig-tail—pound a pig-tail will do, and am
short of shirts. Dear Tom, as for shirts,
ony took 2 whereof one is quite wored out,
and tuther most, but don't forget the pig-
tail, as I ain't had a quid to chor never
since Thursday. Dear Tom—as for the
shirts, your size will do, ony longer. I
liks um long—get one at present, best at
Tower-Hill, and cheap, but be particler to
go to 7 diles for the pig-tail at the Black
Boy, and Dear Tom, acks for pound best
pig-tail, and let it be good—Captain's boy
will put the pig-tail in his pocket, he likes
pig-tail, so ty it up. Dear Tom, shall be up
about Monday, there or thereabouts. Not
so perticler for the shirt, as the present can
be washed, but don't forget the pig-tail
without fail, so am your loving brother
T. P.

P. S.—Don't forget the pig-tail.

The French are not great pipe smokers,
but in Germany the pipe may be said to be
the national utensil. Passing a good part

of his existence at the brasserie, or seated in his arm-chair at the corner of the fire, the German of the old school has adopted a pipe that can remain lighted a long time. The bowl, in porcelain, lends itself easily to ornamentation, while the long stem enables him to hold it without burning his fingers. The student's pipe is shaped something like an Hungarian sabre. It is often the sign of recognition with the initiated ones of the secret societies so prevalent among the German students.

The clay pipe is in general use in Europe : it is made in France, Belgium, Holland, England, Spain, and Italy. The best qualities require considerable care, and the greatest difficulty consists in piercing the stem. The porcelain pipes are preferred in Germany ; they are made of very pure kaolin, and covered with a brilliant enamel. The briarwood pipe is manufactured at Paris and at Saint-Claude, in the Jura. Since Africa has been colonized by the French the red clay pipes, with wide-spreading bowl and Oriental designs, have become familiar. The genuine pipe of this kind is made in Algeria and Morocco, but quantities of imitations are manufactured at Marseilles. The stem for these pipes is usually of cherry or jessamine, covered with its bark. The finest of these stems come from Hungary or from the plateaus of Asia Minor.

The most popular pipe made is the briar-

wood. Some idea of the favor of this wood
is furnished by the fact that some manufacturers make 1500 different styles of briar
pipe, and find a trade for each one of these
styles. These pipes cost from five cents to
twenty-five dollars each, according to design
and the amount of work required to complete them. The most expensive are finished
in meerschaum and amber.

Some of the handsomest pipes in briar
goods are the Pompeiian pipes made by
the famous briar pipemaker, Herr Koch,
of Metz, Germany. The decoration of
these pipes is unusually artistic. To give
an idea, one of the specimens may be described as consisting of a dragon's head.
The skin is black, that hue being produced
by charring the wood with hot irons and
then rubbing it smooth and polishing it ;
each scale is edged with gold. The inside
of the mouth is the natural cedar red
brown of the briar (or more correctly
bruyere) root, and the tongue is of a bright
blood red, this color being the plain wood
highly polished. A large red claw is turned
backward to support the bowl, and the
general appearance of the pipe is handsome
in the extreme. Other fine designs are in
the form of Pompeiian lamps, the bowl
being black at the base, decorated by a cor-
onal of antique pattern, the Mosaic pattern
being produced by the natural wood, the
red polished wood, the charred black, and
the several shades of brown black produced

by skilful charring. One of these pipes is
mounted with three pieces of albatross
quill, laid side by side and connected by
antique silver joints, so that the smoke
travels along each of these before it reaches
the mouth. The stem is really three times
its nominal length. The bowl is perfectly
plain, polished black. On the front, carved
in colors, may be placed the crest, coat of
arms, or monogram of the owner, his club,
regiment, or organization : any of these
designs being executed to order.

When the pipe made its appearance in
France, in the reign of Louis XIV., the
government began to distribute pipes
among the soldiers. Jean Bart was an inveterate smoker, and the story goes that
some Bourbon princesses used to smoke
pipes. There was very little smoking in
Europe in the eighteenth century. No
great man of that time was a smoker.
During the French revolution the pipe was
comparatively unknown. Neither Robes-
pierre nor Danton, nor any one of the leaders of that period, was a smoker. But
when Napoleon's army returned from
Egypt the pipe became fashionable. General Lassalle used to lead his cavalry
charges with a pipe in his mouth ; and
d'Oudinot was the possessor of a splendid
meerschaum, which was presented to him
by Napoleon, and which was ornamented
with stones to the value of about $7500.
General Moreau, when his legs were about

to be amputated, called for his pipe, that
he might smoke it during the operation.

The Restoration brought about a reac-
tion against the pipe, and it was not
until 1830 that it regained a popularity
which it has preserved up to the present
time. Except perhaps in England, the
pipe is considered out of place on the
street ; but at home it is just the thing in
all sorts of society, and it is smoked by
many great men, including Bismarck.

French poets have frequently compared
a man's existence to a lighted pipe, whose
contents pass off in smoke and ashes. In
an old volume of the eighteenth century,
entitled " Morale de Guérard," there is an
engraving representing a young man smoking a clay pipe, and the legend calls him
the " Universal Portrait." This is followed
by a queer old piece of poetry comparing
everybody to a lighted pipe.

From a schism in tobacco-pipes Knicker-
bocker dates the use of parties in the Niew
Nederlandts. The rich and self-impor-
tant burghers who had made their fortunes,
and could afford to be lazy, adhered to the
ancient fashion, and formed a kind of
aristocracy known as the " Long-pipes ;"
while the lower order, adopting the re-
form of William Kieft, as more convenient
in their handicraft employments, were
branded with the plebeian name of " short-
pipes."

VII.

Origin of the Cigarette — Famous
WomenSmokers—Empresses, Queens,
Princesses, and Women of Fash-
ion who Find Comfort in the Dainty
Cigarette or the Fragrant Cigar—
In the Orient the Fair Ones Gossip,
Smoke, and Drink Coffee—A Peep
into the Harem—Curious Riddles
and Proverbs about Tobacco.

What is the origin of the cigarette?
Some authorities assert that its use is as old
as that of smoking tobacco itself ; but it is
not possible to be very precise on this point,
although it is probable that the earliest
smokers made their own cigarettes.   The
great vogue of the cigarette in France dates
from the same time as the introduction of
the cigar, namely, 1830, when the influence
of romanticism contributed powerfully to
making it fashionable.   Hugo's drama of
"Hernani" started the taste for Spanish
things, and the young literary men and
artists at once took to smoking cigarettes.
In Spain everybody smokes, and generally
smokes the cigarette.   As soon as two men
meet one offers cigarettes and the other the
light.   The Spaniards smoke slowly, and
do not light a second cigarette immediately
after the first one is finished.   Sober in all
things, the Spaniard knows how to enjoy
the final taste and the souvenir.   He smokes
everywhere and at all hours : before his re-

pasts, afterward, and between meals. At
the theatre, at the bull fights, in the offices,
and on the railways tobacco has conquered
its citizenship.

The Spaniards have a proverb to this
effect : " A paper cigarette, a glass of fresh
water, and the kiss of a pretty girl will
sustain a man for a day without eating."

One of the greatest charms for the smoker is to make the cigarette himself, and the
Spaniards are exceedingly clever at this
work. The most remarkable quality of
their sovereign, Ferdinand VII., was that
he could make two cigarettes simultane-
ously, one with each hand.

The manufacture of cigarettes in France
began in 1843, and at first the factory at
Gros Caillou was able to supply the demand. Now there are seven factories at
work, employing 2,000 women, who turn
out 400,000,000 cigarettes every year.
French cigarettes are made of ordinary and
superior scaferlati tobacco and sold in various colored packages of twenty ; the
prices range from six cents to forty cents a
package. The foreign cigarettes on sale in
France are generally made of Levant
tobacco, and American brands are also
obtainable in a few places. The French
cigarettes are manufactured by a machine
that fabricates 15,000 in ten hours.

A vast majority of the Empresses,
Queens, and Princesses of the world rest in
the conviction that life bears a more beau-

tiful aspect when seen through the opalescent clouds of fragrant smoke that issue from their delicate mouths.

Empress Elizabeth, of Austria, smokes from thirty to forty Turkish and Russian cigarettes a day, and for many years it has been her inveterate custom to puff away after dinner at a strong Italian cigar, one of those with a straw running through it, and which is brought to her with her cup of Turkish coffee every evening on a gold salver. She says herself that smoking soothes her nerves, and that whenever she feels "blue" a cigar or a cigarette will do more than anything else to cause her to see things in a happier light. She is a perfect Greek and Latin scholar, and when writing she smokes almost continually. On her writing table are always a large silver box of repousse work filled with cigarettes, a match-box of carved Chinese Jade, and a capacious ash receiver, made of the hoof of a favorite hunter, which broke its spine over a blackthorn hedge. Almost mechanically Her Majesty lights cigarette after cigarette, as she sits in her great writing-room, which is fitted up with carved-oak panels and Gobelin tapestries, the sombre hue of the walls being relieved here and there by trophies of the chase.

The Czarina of Russia, who is likewise one of the vassals of King Nicotine, smokes in a somewhat more indolent and almost Oriental fashion. Stretched on the

silken cushions of a broad low divan, at
Gatchnia, she follows dreamily with her
beautiful dark eyes the rings of blue smoke
that her crimson lips part to send upward
into the perfumed air of her boudoir, a
boudoir which she calls her "den," and
which is copied from one of the loveliest
rooms of the Alhambra.

Queen Marguerite, of Italy, is another
of the royal ladies who see no harm in the
use of tobacco. Her flashing black eyes
look laughingly through fragrant clouds
of smoke, and she is wont to declare that
her cigarette is more essential to her comfort than anything else in life.

Christina, Queen Regent of Spain, is a
great advocate of tobacco. She consumes
a large quantity of Egyptian cigarettes,
and there is nothing that her little
"Bubi," His Most Catholic Majesty King
Alphonso XIII., enjoys more than when
his mother permits him to strike a match
and apply the flame to the end of her
cigarette. When thus engaged the little
fellow laughs merrily, and indulges in all
sorts of antics, like a light-hearted little
monarch that he is.

His Holiness Pope Leo XIII. at any rate
does not consider the use of tobacco as a
vice, else he would scarcely have conferred
the Golden Rose on so inveterate and confirmed votaries of the weed as Queen Christina and the ex-Crown Princess of Brazil.
Indeed, there is every reason to believe

that, like many other enlightened spirits, he regards the objection to cigarettes as being mere smoke after all.

The smoking paraphernalia of the beautiful and voluptuous-looking ex-Queen Natalie, of Servia, is of the most elaborate and magnificent description, while the poet-Queen of Roumania, so well known in the literary world under the pseudonym of " Carmen Sylva," is content with a gold cigarette case suspended to her chatelaine.

The Comtesse de Paris, the Queen de jure of France. is addicted to mild Havanas of delicious flavor, and her daughter, Queen Amelia, of Portugal, is a source of considerable fortune to the manufacturers of Russian cigarettes at Dresden. All the Russian Grand Duchesses and most of the imperial Archduchesses of Austria, including Marie Therese, Elizabeth and Clothilde, smoke to their hearts' content and in the most public manner, and their example is followed by Queen Olga, of Wurtemberg, who is a daughter of Czar Nicholas ; by Queen Olga, of Greece, who is likewise a Russian Grand Duchess ; by the Princesses Leopold and Luitpold, of Bavaria, and by Queen Henrietta of Belgium.

Queen Victoria has an intense horror of smoking, and it is strictly prohibited at Windsor Castle, at Balmoral, and at Osborne. This, indeed, is one of the main reasons why the visits of the Prince of Wales to his august mother are so brief,

and so few and far between, for the heir
apparent to the English throne is so little
accustomed to self-denial and so fond of
smoking, that he is scarcely ever to be seen
for an hour together without a cigar or
cigarette between his lips. The Princess
Louise, Marchioness of Lorne, smokes, but
both his wife and his daughters, especially
Princess Maud, are accustomed to indulge
in a cigarette when in their morning-room
at Sandringham or Marlborough House.
Many, in fact most of the great ladies of
France, such as the Duchesses de Mouchy,
de la Rochefoucauld-Doudearville, d'Uzes,
and de Maille, are fond of cigarettes, the
fashion having been set in France some
five and thirty years ago by Empress
Eugenie, who, like all Spaniards, was
never at her ease except when puffing
clouds of fragrant smoke from her lips.
Indeed, during the Napoleonic régime
there was scarcely a corner in the palace
of the Tuileries, St. Cloud, or Compiegne
which was not redolent with the fumes of
tobacco. Of the members of the Imperial
French Court, Napoleon's cousin, Princess
Mathilde, the Princesses de Sagan, the
Duchess de Persigby, the Marquise de
Gallifet, the Marquise de Beibœuf, and the
Comtesse de Pourtales, may every one of
them have been said to have seen life only
through hazy clouds of smoke. In Austria and Hungary all the great ladies divide
their loyalty equally between their beloved

Emperor on the one hand and King Nicotine on the other, and many is the time that
the Princess Metternich, Princess Leon-
tine Furstenberg, Margravin Pallavicini,
Countess Shonborn, Princess Clam-Gallas,
and Countess Andrassy have been seen
smoking on the race-course of the Freude-
nau, or even in the Stadt Park, while listen-
ing to the strains of Strauss's orchestra.

It is in the Orient, however, that smoking has been developed into a fine art.
Debarred from all the social pleasures and
active mode of life of their European sis-
ters, the ladies of the Zenana are restricted
to gossip, coffee, and tobacco. Nowhere
else in the world are these three things
brought to such a standard of perfection.
A fair idea of the importance attached
thereto by Turkish women of high rank
may be obtained by a visit to the Harem
of the Khedive of Egypt at the Ismailia
Palace on the banks of the Nile. The
audience chamber of His Highness's only
wife is a casket fit for a jewel. The furniture is of ivory and mother-of-pearl, and
the hangings of silvery satins, embroidered
with pale roses and violets in silk and silver thread. The ceiling and woodwork are
painted with groups of flowers, and the glass
in the windows is milk-white, while the
floor is covered with thick white Aubusson
rugs, strewn with a design of rose leaves
and buds. Here, lying back on a low vel-
vet divan, is the Vice-Queen, smiling her

welcome to the approaching visitor. She
ts stiil extremely beautiful, although a lit-
lle too short. Her face is brilliant and
lovely like a Titian or a Rubens ; her eyes
are very large and velvety, full of the
slumberous fires of the Orient ; her scarlet
lips are like a double camelia petal, and
her skin of the warm, creamy whitness of
the tea-rose. She is generally clothed in
white silken tissues, cut à l'Europienne,
with a great profusion of marvellous lace,
and a perfect shower of pearls and diamonds glittering on her hair, on her white
bosom, encircling her wrists and covering
her small, plump hands. Diamonds
sparkle everywhere ; the tobacco-box,
which lies on a low inlaid table near the
Vice-Queen, is studded with them. The
inkstand and penholder which adorn her
writing-desk are all ablaze with splendid
gems. Her Highness's slippers are thickly
sewn with brilliants, and more jewels form
monograms on all the dainty trinkets which
surround her, from her gold footstool to her
powder-box and tortoise-shell hand-glass.
On her heart the Vice-Queen wears a min-
iature of her husband framed with huge
diamonds and rubies, and around her waist
is a broad band of the same stones to which
is suspended a fan of snowy ostrich feath-
ers, its handle encrusted with pearls, emer-
alds, and sapphires. In spite of all this profusion of jewelry, there is nothing discord-
ant in the sovereign's appearance. The

nature of the luxury is in perfect keeping
with her Oriental style of beauty, and the
setting in absolute harmony with the great
brilliancy of the picture she presents.

The Vice-Queen frequently smokes a
narghile (water-pipe). This suits her
style of beauty even better than the more
prosaic cigarette. The bowl is of engraved
rock crystal mounted in chased gold, fashioned in the form of a lotus flower. The
tube is concealed by a deftly wrought net-
work of pink silk and gold thread, while
the amber mouth-piece and gold plateau
are one mass of sparkling jewels.

Here are some curious riddles and
proverbs about tobacco. A little book
hawked about the country in the reign of
Queen Anne contains the following riddle :

> " What tho' I have a nauseous breath,
>   Yet many a one will me commend ;
> I am beloved after death,
>   And serviceable to my friend,"

to which is appended the answer, " This is
tobacco, after cut and dryed, being dead,
becometh serviceable." The following
" quaint conceit" is still more clever :

> " To three-fourths of a cross add a circle complete ;
>   Let two semicircles a perpendicular meet ;
>   Next add a triangle that stands on two feet,
> Then two semicircles and a circle complete."

" To three-fourths of a cross add a circle complete :	TO
Let two semicircles a perpendicular meet :	B
Next add a triangle that stands on two feet :	A
Then two semicircles and a circle complete :	CCO

A cutty bowl, like a Creole's eye, is most prized when blackest.

Coffee without tobacco is meat without salt.—*Persian.*

A wealthy Englishman, who left his estates to Lord Chatham, in admiration of his talents, possessed a tobacco-box, on which, under a skull, was engraved a Latin quotation, which has been thus rendered in English :

' Of lordly man, how humble is the type,
   A fleeting shadow, a tobacco-pipe !
 His mind the fire, his frame the tube of clay,
 His breath the smoke so idly puffed away,
 His food the herb that fills the hollow bowl,
 Death is the stopper, ashes end the whole."

Senator Thurman and Senator Edmunds, between whom there was a strong friendship, were lovers of the pipe. Albert Pike, the poet, has written of them :

" Not from cigars these Senate stars
    Their inspiration drew ;
  Old pipes they smoked, as they sat and joked—
    Yes, pipes, and cob pipes, too !"

THE JOKER'S DICTIONARY.

Thousands of men, when in the society of ladies or gentlemen, want to be entertaining and amusing companions, but too often find that they cannot.

This great and original book comes to the aid of just such people ; it does for the man who wants to be witty what Webster's Dictionary does for the man who wants to be wise in the use of words.

It is a perfect cyclopedia of wit and humor.

It contains 326 pages, six illustrations, and is arranged according to subject, alphabetically. That is to say, it is in the style of a dictionary. You can find Jokes, Stories, and clever bits of repartee, brilliant jests and flashes of merriment, on almost every subject likely to come up in social intercourse.

Price, **25 Cents**, post paid.

www.ingramcontent.com/pod-product-compliance
Lightning Source LLC
Chambersburg PA
CBHW030541270326
41927CB00008B/1473